THE MAN WHO LOST HIS HEAD

Douwe Draaisma

The Man Who Lost His Head

On Illusions and Delusions of the Mind

Translated by Jane Hedley-Prôle

REAKTION BOOKS

Published by
Reaktion Books Ltd
2–4 Sebastian Street
London EC1V 0HE, UK
www.reaktionbooks.co.uk

Third edition © 2022 Douwe Draaisma/Groningen
Originally published by Historische Uitgeverij, Groningen
English-language translation © Reaktion Books 2025

This book was published with the support of the
Dutch Foundation for Literature

N ederlands
letterenfonds
dutch foundation
for literature

EU GPSR Authorised Representative
Logos Europe, 9 rue Nicolas Poussin, 17000, La Rochelle, France
email: contact@logoseurope.eu

Printed and bound in Great Britain by Bell & Bain, Glasgow

A catalogue record for this book is available from the British Library

ISBN 978 1 83639 088 6

CONTENTS

Through the Looking-Glass 7

ONE

The Three Christs of Ypsilanti 23

TWO

Living in the Knowledge of Being Dead 65

THREE

The Murder of the Widow Van Sandbrink 93

FOUR

Phantoms and Illusions 107

FIVE

Whole at Last 148

SIX

Grief Hallucinations 171

REFERENCES 191
ACKNOWLEDGEMENTS 203
PHOTO ACKNOWLEDGEMENTS 205

Through the Looking-Glass

The guillotine had two fathers. Both were physicians. Antoine Louis, credited with the first design, was a renowned surgeon and secretary of the Royal Academy of Surgery in Paris. He engaged Tobias Schmidt, a Prussian harpsichord maker, to build a prototype. Schmidt undertook to do so for the modest sum of 960 livres, requesting to be granted the patent as compensation. The French Ministry of the Interior turned his application down. Herr Schmidt had admittedly constructed 'a useful invention of a lethal kind' but, since it was only suitable for carrying out sentences, he would have to relinquish the patent to the government.[1]

The second father was Joseph-Ignace Guillotin. Favouring a more humane method of execution, he had already proposed using a decapitation device back in 1789, the first year of the French Revolution, but the timing was unfortunate. The revolutionaries – including Robespierre, later architect of the Reign of Terror – were debating an even more humane option: abolishing the death penalty altogether. Robespierre was in favour.

It wasn't until nearly three years later, early on the morning of 17 April 1792, that the first experiments took place. The apparatus had been erected in the courtyard of the Bicêtre Hospital, just outside Paris. Guillotin and several of his colleagues looked on as a couple of live sheep were successfully decapitated, followed by three human cadavers.

That the guillotine, as the machine came to be known, was invented by doctors – and not by an executioner wanting to mechanize his handiwork – was no coincidence. Their intentions were democratic and humane. By introducing a uniform method of execution for all condemned criminals, they sought to end what amounted to class justice in extremis. Arsonists were burnt at the stake. Thieves were hanged. Only aristocrats had the privilege of beheading by the sword. A week after the experiments on sheep and corpses, a highwayman by the name of Pelletier was executed in a way that would previously have been considered far above his station. The execution was performed with clinical efficiency, somewhat to the disappointment of the crowd who – at 3.30 a.m. – had gathered to watch. They had hoped for greater spectacle. From Pelletier's death onwards, the guillotine would serve the revolutionary ideal of equality in exemplary manner for nearly two centuries. Every French *département* received a blueprint of the design.

Besides egalitarian, the machine's designers lauded it as humane: death resulted swiftly, if not instantaneously, so that suffering was not needlessly protracted. The design was functional: a rope, a pulley and a diagonal blade suspended between two upright posts, which shot straight down, with virtually no resistance, until it hit the neck. The guillotine never failed. However questionable a trial, however mendacious the witnesses, however dubious the evidence, once on the scaffold, a sharp blade and gravity ensured that sentence was always carried out flawlessly in a fraction of a second.

Guillotines appeared in prison courtyards all over France. But in its glory days, the device was a public – not to say theatrical – object. During the Reign of Terror, from around the spring of 1793 to August 1794, on the Place de la Révolution and dozens of squares elsewhere in France, it was given a stage and an audience. The executioner who operated the guillotine became a mere extra. The machine itself, towering far above the watching crowd, claimed the starring role. Executions became shows. And there were very many of them. In Paris, around 3,000 people were guillotined, and in France as a whole around 17,000 – always in front of a crowd of onlookers. At the height of the Terror, shortly before the fall of Robespierre, 68 heads rolled in Paris in a single day. Robespierre himself was one of the last victims, his earlier proposal to abolish the death penalty having not been taken up. At the end of the Reign of Terror, the guillotine was removed from the stage and taken indoors.

The French long remained faithful to the mechanical executioner. The last head, that of Hamida Djandoubi, an agricultural labourer convicted of torture and murder, rolled on the morning of 10 September 1977, three months after President Valéry Giscard d'Estaing had refused to issue a pardon. Under Giscard's successor, François Mitterrand, the death penalty was abolished.

⁂

While the Terror raged outside, a Parisian silversmith and watchmaker sat in his workplace, busy with a quite different invention. By the end of the eighteenth century, timepieces had become so refined that table clocks and watches could run for several days after a single winding. The watchmaker had finessed this yet further, constructing devices with mechanisms that functioned almost without friction. But almost was not the same as completely. Could he not, he wondered, with a bit more effort and skill, construct a machine that, once started, would continue to run forever? Why should he not be the man to achieve the dream of *perpetuum mobile*? This vision seized him – he would astonish the world with his perpetual motion machine.

Seemingly blind to the horrors going on around him, the watchmaker set to work, grinding and filing cogwheels, shafts and springs. Gradually, his labours became obsessive. He scarcely allowed himself time to eat. Exhaustion and lack of sleep began to take their toll. Eventually the poor man showed signs of derangement, and his family decided to seek help. He was placed in an asylum at the Bicêtre, the same hospital where, a few years earlier, the guillotine had been tested. There the watchmaker was placed under the care of Dr Jean-Baptiste Pussin. Over the course of 1800 and 1801, Pussin's superior, Philippe Pinel, would record his case history.

The watchmaker might be manic – he sang, cried or danced incessantly – but he was not dangerous and, after a while, he was permitted to wander freely around the asylum. Still obsessed by the idea of perpetual motion, he sketched one design after another on walls and doors with a piece of chalk. What worried Pussin, though, was that he was afflicted by a new symptom.

A terrible thing had happened to him, the watchmaker told anyone who was prepared to listen. He claimed that he had lost his head on the scaffold. After being condemned to death, he had been made to kneel under the guillotine and, when the blade fell, his head had been severed from his body. The executioner's henchman had tossed the head among those of many other victims. He would now be dead, he explained, had the judges not regretted this cruel punishment and given orders to replace the severed heads on the decapitated bodies. Unfortunately, the men entrusted with this task carried it out sloppily. When it was his turn, they had randomly grabbed a head from the pile: the wrong one. And now he was having to go about with someone else's head, which – to make matters worse – was considerably inferior to his own. For example, he himself had had excellent teeth, he told his fellow inmates, whereas the ones he had now were rotten and decayed. Nor was he happy about the alien head's coiffure: 'What a difference between this hair and that of my own head!'[2]

The delusion proved extremely persistent. There was no arguing against it. He was blind to the illogicality of being able to remember his own head with someone else's. While this illusion persisted, he was clearly not fit to be discharged. Finally, Pussin resorted to a ruse. The saints venerated during the Middle Ages included 'cephalophores', or head-carriers: martyrs who had been beheaded for their Christian faith but who had subsequently picked up their heads and gone on their way again, holding them in their hands or tucked under their arms. More than 130 of these legends were recorded.[3] In the third century AD, for instance, the nine-year-old Justus of Beauvais had been decapitated by Roman soldiers after admitting to being a Christian.[4] When the severed head remained alive – and started to pray into the bargain – the soldiers panicked and ran away. His father found him sitting with his head in his lap. The boy asked for his head to be taken to his mother, so that she might kiss it before he was buried. But the most famous head-carrier was St Denis, a Paris bishop also said to have been beheaded for his faith. After being executed in Montmartre, the saint picked up his head, kissed it and walked 11 kilometres (7 mi.) to the place where he wished to be buried, preaching as he did so. Any Parisian with even a smattering of education was familiar with the 'miracle of St Denis'. A sculpture depicting the saint with his head in his hands adorns the left-hand portal of Notre Dame; the watchmaker must almost certainly have seen it at some point. Pussin confided his plan to a patient who had just been declared cured and gave him careful instructions. Then he arranged for a meeting with the watchmaker. The accomplice cleverly turned the subject to the miracle of St Denis, casting doubt on its veracity. The innocent watchmaker took the bait. The story of the saint was true, he maintained. You *could* live after having lost your head. After all, it had happened to him, hadn't it? The accomplice seized his chance. 'Fool! How could St Denis kiss his own head?' he retorted. 'Was it with his heels?'[5] Laughter broke out at the watchmaker's expense. He seemed to realize with a shock that what he believed was in fact impossible and withdrew, confused. He never spoke of the swapped heads again. Not much later, he was declared cured and discharged.

※

Delusions are windows into their time. For decades after the French Revolution, French asylums treated people who had been unhinged by its turbulence and terrors. In 1801 Pinel listed various 'exciting causes' – or triggers, as we would call them now – of mental derangement in 113 psychiatric patients. The biggest category was 'domestic misfortunes': divorce, bankruptcy or the loss of a child, accounting for 34 cases. But it was immediately followed by the category 'events connected with the Revolution': thirty cases. Many delusions spun themselves round the guillotine.

Pinel knew from personal experience how traumatic it could be to witness an execution. He had escorted the carriage that conveyed Louis XVI to the guillotine and found himself where he really did not want to be: right at the front, at the foot of the scaffold. That same day he wrote to his brother: 'To my great regret I was obliged to attend the execution, alongside other citizens of my district. I write with a heart heavy with sorrow, in a stupor of profound consternation.'[6] By then he was 47 and, as a doctor, had seen his fair share of suffering and death, yet the brutality of the execution had thrown him off balance.

Anyone who wisely tried to protect their psyche by steering clear of executions still risked being confronted by the many gruesome depictions of them. Illustrators and political cartoonists liked to show the moment when blood spouted from the neck or the executioner triumphantly held up a dripping head by the hair in front of the mob. A macabre cartoon from 1793 by the engraver

The execution of Louis XVI on 21 January 1793. Philippe Pinel was designated as a member of the armed escort, and to his dismay was forced to witness the execution from close by.

Villeneuve shows the arrival in the underworld of Louis Capet, the former Louis XVI. Head under his arm, he has just stepped out of Charon's boat to be met by a welcoming committee of previously guillotined aristocrats, either kneeling and holding out their heads to him, or clutching pikes on which their heads are impaled.

And then there were the stories, from which in Paris – a hotbed of news and rumour – there was absolutely no escaping. After the execution of Charlotte Corday, condemned for the murder of Jean-Paul Marat, a revolutionary had retrieved her head from the basket and slapped it in the face. A blush of indignation was said to have appeared on her cheeks. Doctors debated whether consciousness really did expire directly after decapitation. Some believed that thoughts continued to circulate in the head for a

Arrival of the decapitated Louis XVI in the underworld, engraving by Villeneuve, c. 1793. The accompanying text states that he was welcomed by 'a great company of villains, former crowned heads'.

little while but could no longer be expressed, due to lack of breath. Stories were spread about severed heads that had been seen to grimace or grind their teeth in the basket.

While by day the 'national razor' steadily continued to do its work, cheered by a baying mob, at night it was a haunting presence, spawning nightmares and terrifying visions. Sometimes fear of the guillotine was itself lethal. Continual fear of death, Pinel wrote, puts the body in a permanent state of consternation, causing some to pine away. In his own Bicêtre, he had overseen the care of two Austrian prisoners of war whose fear of ending on the guillotine was so all-consuming that they eventually died. The archive of the Charenton lunatic asylum records the case of Jean-Pierre Laujon who, after initially fleeing to England and then returning to France to fight the revolutionaries, had been taken prisoner. He escaped the guillotine on grounds of insanity: he claimed that he had already been beheaded, and that his real head was still in England. Just like the watchmaker, he believed that he was now walking around with someone else's head. In 1829, 35 years after the end of the Reign of Terror, Laujon was still convinced that he 'had been put to death and then brought back to life'.[7] Much later still, in 1855, when Pinel was long dead, one of his successors in the Bicêtre admitted a man who declared that his father had died on the guillotine and that he was related to Louis XVI.[8] The guillotine was still able to claim victims even a generation later.

※

Delusions are time-bound. Their expression is tied to a specific segment of time, place and personal history. In 1838 Pinel's student Jean-Étienne Dominique Esquirol claimed that he could have written the revolutionary history of France, from the storming of the Bastille up to and including Napoleon, on the basis of admissions to lunatic asylums. When he made this claim, Napoleon had been dead for seventeen years. Had Esquirol lived just three days longer, though, he would have seen the emperor's body being conveyed to Les Invalides for interment on 15 December 1840. What followed would show how right Esquirol had been. Tens of thousands jostled for a spot along the route taken by the funeral carriage, keen to pay their respects. The run-up to the ceremony – from the exhumation of the miraculously well-preserved body on St Helena to the triumphal journey through France from the port of Cherbourg to Paris – had caused a great stir. That same year alone, thirteen or fourteen 'Napoleons' were admitted to the Bicêtre.[9] On top of that, the asylum found itself taking in various 'sons' and 'wives' of Napoleon. Napoleon might have been dead for almost twenty years by then but, as the 'exciting cause' of delusions of grandeur, he appeared eternal.

Yet at the same time, delusions are timeless, since they have certain set characteristics. Even in their most bizarre and incomprehensible manifestations, there is still an order, a system. That system expresses itself in different ways, whether through painstaking efforts to maintain consistency, taking absurd premises to logical conclusions or the rationality with which patients explain their own actions and perceptions to others. For the patient themselves, delusions are more a hypothesis, a scenario, an explanation – something that helps them make sense of an intensely confusing experience. The watchmaker might have looked in the mirror and failed to recognize himself. In his recollection he looked very different, with teeth that were good, not rotten. This face in the mirror had to be someone else's; it certainly wasn't his. But how on earth could someone else's head have ended up on his shoulders? And how had he lost his own? In his confused state, the images of all those heads rolling under the guillotine – perhaps witnessed in real life, seen in political cartoons or heard about – presented the solution. He must have been executed: *that* was how he'd lost his head! In the chaos that followed, someone had apparently put the wrong head on his body. That must have been how it happened, how else? To the watchmaker, the switching of heads wasn't a delusion but the most logical explanation for the shock of being confronted by a ravaged face in the mirror.

It is often touching to see how people whose mental balance has been disturbed try to keep a grasp on destabilizing experiences and thoughts by applying logic and citing facts. Laujon, the patient in Charenton who thought he had died and subsequently been brought back to life, would in our time probably be diagnosed with Cotard's syndrome. Patients with this condition believe that they are already dead or that their body lacks vital organs like the heart or lungs. The first 'official' case was presented by the Parisian psychiatrist Jules Cotard in a clinical lesson in 1880. People with this syndrome think that they are walking around as hollow corpses – as the living dead – and nothing will persuade them otherwise. However preposterous it might seem to believe that you are dead, patients with this delusion nevertheless draw logical conclusions (see Chapter Two). They refuse to eat: after all, a dead body doesn't require food. Some try to set fire to their bodies: a putrefying corpse must smell appalling; high time it was cremated. Others have a history of suicide attempts, which they abandon once in the grip of this delusion: they are already dead, right? In 1835 Lucas Kier, a tobacco grower from Amersfoort, committed a murder while suffering from Cotard's syndrome (see Chapter Three). In a farewell letter he stated, with references to the Book of Revelation, that he had been killed a year earlier, during the Last Judgement, but was now in the state of 'Second Death'. That explained how he could have died and yet still live.

Another constant factor in delusions is the response they prompt among a patient's family and friends and often their doctors. It is difficult for people close to the patient to accept the situation. Their impulse is to challenge delusions, disprove them and show their absurdity. People with Capgras syndrome – also discussed in Chapter Two – suffer from the delusion that their loved ones have been replaced by a body double. They are utterly convinced that their real wife or daughter has gone, that the woman who comes to visit and looks just like her must therefore be an imposter. The person whose identity is doubted has the natural urge to prove they're not an imposter, for example by telling the patient something that only they could both know. But that never makes the delusion go away. On the contrary. Instead, the patient, already suspicious, gets even more paranoid. It's worse than they thought: apparently their vanished loved ones are also in on the plot – how could the imposter otherwise have access to this knowledge? Here, again, however faulty the conclusion, the logic is infallible.

The attempt to free the watchmaker from the grip of his delusion by getting him to see its absurdity drew on a long tradition, even back in 1801, and it's an approach that has also been tried in the more recent history of psychiatry. Chapter One recounts a two-year experiment to 'cure' three schizophrenic men who each believed they were Jesus Christ by putting them all in the same psychiatric hospital and confronting them with each other. The social psychologist Milton Rokeach started this ambitious project

– which came to be known as 'the three Christs of Ypsilanti' after the Michigan state hospital where the experiment took place – in 1958. It explored the resilience and solidity of delusions, and their significance for the patients. At the same time, Rokeach's case study is a depressing account of manipulation and deception, dating from an era when you could do whatever you liked to psychiatric patients in the name of science.

❈

You don't need to have mental-health issues to experience sensations that force themselves upon you, sometimes overwhelmingly, though they have no basis in reality. People who lose a limb, for example, can still sometimes 'feel' that missing body part. These 'phantom sensations' (see Chapter Four) aren't confined to missing limbs, either – cases have also been recorded of phantom penises (in the case of people who have had gender reassignment surgery), phantom sounds (in deaf people), phantom images (in blind people) and phantom smells.

Phantom sensations are often described in the same terms as delusions: they are unreal, illusory, deceptive, imaginary. The difference has to do with insight: in psychiatry, a delusion is by definition a belief that a patient perceives as compelling and unquestionable. The pain that someone can experience in a missing arm after amputation feels all too real, but at the same time they know their arm isn't there any more. Phantom sensations raise the question of where the limits of the body actually are. Is your body what you see in the mirror, or is it the sum of your bodily sensations? If someone feels that their amputated arm is sticking out to the side, and for years walks through doorways sideways to avoid banging it, does that arm still belong to the body or not?

By contrast, people who suffer from what's called body integrity identity disorder perceive the contours of their bodies as narrower than they are in reality. To them, a foot, a lower leg or an arm has felt 'alien' from an early age, as if it doesn't really belong to their body (see Chapter Five). As they grow up, that sense of alienation increases, finally becoming so powerful that they crave to have the body part amputated. Since very few doctors will cooperate in such cases, patients who can afford it sometimes turn to an illegal circuit of surgeons willing to carry out the operation. The disorder mirrors phantom sensations: a phantom arm has objectively disappeared, but it is still perceived as part of the self; the 'alien' arm is there, but it is not perceived as part of the person's own body.

To an outsider, trying to shield a missing arm from knocks or wanting to get rid of a healthy arm seems every bit as incomprehensible as a delusion. However, just as in the case of delusions, these feelings are perceived with great subjective conviction and are a reaction of the self – or the brain – to sensations that are all too real. An illusory sensation is still just that: a sensation.

※

In 1971 the Welsh GP William Dewi Rees began to investigate a phenomenon he had noticed in some of his older patients who had recently been widowed (see Chapter Six). They told him that they still occasionally saw or heard their loved one: sitting in their usual chair, pottering about upstairs, humming in an adjacent room. Such 'hallucinations of widowhood', as Rees initially called them, were far from uncommon: just more than half of the three hundred respondents had experienced them. These days, the phenomenon has been renamed 'grief hallucinations', a more appropriate term – not just because widowers have them slightly more often than widows, but because they can follow the loss of someone other than a life partner. Although grief hallucinations are common, they are rarely talked about, especially by those who experience them: sometimes because of misplaced shame or fear of being thought crazy, sometimes because it is seen as an experience too intimate to be shared. Nearly all such hallucinations are perceived as reassuring, as giving a sense of presence, however temporarily. The question of course is: where do they come from? What causes them? Or should we ask: why do they occur?

※

In *Through the Looking-Glass* (1871) by Lewis Carroll, Alice, curious about the world on the other side of the mirror above the fireplace, climbs up onto the mantelpiece and steps through the mirror. On the other side she finds herself in a fantastical world where you have to run to stay in the same place, plants can talk and chess pieces come to life. Halfway through the book she gets into a conversation with the White Queen. Alice, the voice of common sense, claims that you can't believe impossible things. The queen is having none of it. It's just a question of practice, she says sternly, adding that, when she was younger, she sometimes 'believed as many as six impossible things before breakfast'.[10]

As a result of their profession, psychiatrists, neurologists and clinical psychologists spend part of their time through the looking-glass. Their patients believe the most impossible things. That they are made of glass. That they are holding a cup in the hand of their amputated arm. That they drowned two years previously. That the CIA is spying on them through electrical sockets. That they are kidnapped by aliens at night. That a member of the royal family is in love with them. That they are Jesus Christ. The variations are endless – or so it would seem. In reality, however, delusions and phantoms, illusions and hallucinations form a mosaic in which patterns can be recognized. Some are linked to specific organic injuries, most often brain injury, thus helping to identify the neurological circuits crucial to retaining mental balance. Others are an attempt to wrestle intensely confusing sensations into a context that makes sense. What they have in common is the desperate ingenuity with which the self tries to preserve coherence and cohesion – an inner order – as if reason were the very last thing a confused soul wishes to relinquish.

ONE

The Three Christs
of Ypsilanti

When it opened in 1931, the Ypsilanti State Hospital in Michigan seemed less like an institution and more like a town – and a decent-sized one at that. It had its own water tower and electricity plant, and was virtually self-sufficient. The patients worked in greenhouses, allotments and orchards, milked cows, carried out repairs in the carpentry workshop and went to church in the hospital's own chapel. In the spring of 1937, the local *Ann Arbor News* published a report on life in the institution, featuring photos of the beauty salon for female patients, the theatre, the gymnasium, the dentist at work, the airy dormitories, the kitchens, canteens and laundries, the laboratory and the operating theatre.[1] Not a straitjacket in sight, the visiting reporter noted.

In 1938 the psychiatrist Orus R. Yoder was appointed as the hospital's medical superintendent.[2] Where possible, he promoted a humane programme of therapy. More than half the patients had been diagnosed with schizophrenia. Their tendency to retreat into a private world of delusions was countered through all kinds of occupational therapy, alongside work in the kitchens, bakeries and laundries. But Ypsilanti also had its own choir, orchestra and newspaper. The management experimented with therapies that,

Ypsilanti had its own weaving mill where patients made curtains,
cushion covers and rugs by way of occupational therapy, photograph
from *Ann Arbor News* (11 March 1937).

back then, were progressive, involving music, sports, amateur
dramatics and painting.

In 1931 Ypsilanti was designed to hold around 1,000 beds.
Within the space of a decade, though, Yoder and his staff found
themselves entrusted with the care of more than 4,000 patients.
Sent by family doctors, judges and desperate relatives, a never-
ending stream of disturbed minds flowed into Ypsilanti: manic-
depressives, alcoholics, the feebleminded, epileptics, people with
dementia, brain damage, Korsakov or syphilis. They spilled into
the overcrowded corridors and wards. For many, smoking was
their only diversion and, since their allowances could be spent on
little else, the wards were blue with smoke all day long. Patients
rarely got to see one of the hospital's four or five psychiatrists – at
most once a year. In the dormitories – which already provided

very little privacy – the chair that had initially been placed next to each bed was removed, and the beds were pushed together until they were virtually touching.

The statistics tell a different, much grimmer story than the upbeat photos in the newspaper. An average annual report of the mid-1950s records 760 new intakes. Fewer than three hundred people were discharged as cured. In that same year, 213 had died in the institution, some as a result of suicide or an attack by a fellow patient. Another 71 had escaped. Those who were reasonably mentally stable could work in the greenhouses or kitchens but, for most patients, the days in Ypsilanti were long, monotonous and dangerous.

Perhaps the biggest danger was posed by the doctors and nurses themselves. Those 213 deaths included patients who had

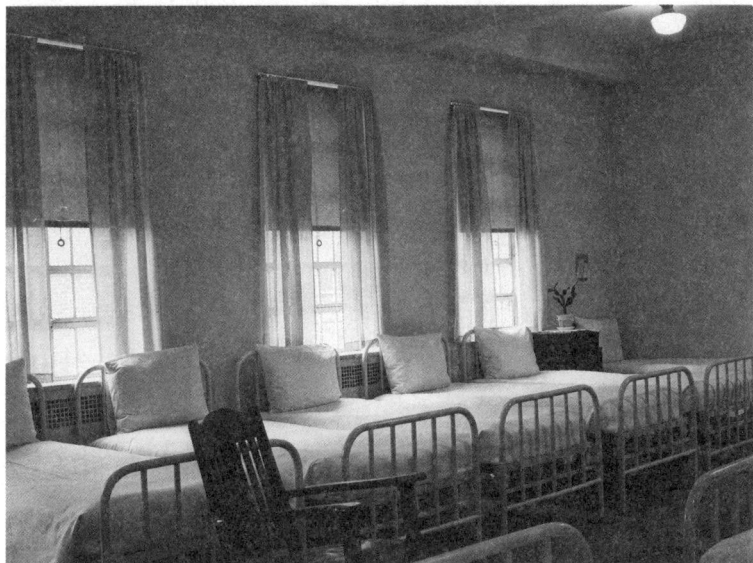

Dormitory in Ypsilanti, from *Ann Arbor News* (19 February 1937). According to the caption, the six-year-old hospital was already so overpopulated that the regular standards of comfort and hygiene had had to be 'temporarily abandoned'. Temporarily turned out to mean permanently.

expired as a result of undergoing therapy. Yoder was always keen to try the latest medical interventions. Like lobotomies, where a surgeon drove a small chisel through a patient's eye socket into their brain, then swished it back and forth three or four times, like a windscreen wiper, to sever the connections between the prefrontal lobe and other parts of the brain. (Some of the earliest films showing how to perform lobotomies were shot in Ypsilanti.) Another therapy targeted patients in the last stage of syphilis. Heavy metals were mixed into their food, the theory being that their bodily reaction to the poison would reverse the infection. Syphilis sufferers were also deliberately infected with malaria, as it was thought that high fever would kill off the microorganisms that caused syphilis. For depressed, suicidal patients, electro-shocks were the standard treatment. In the belief that schizophrenia and epilepsy cancelled each other out, schizophrenic patients were given medication that caused fits – in some cases so severe that their backs were broken.

In those days, psychiatric patients were fair game. You could do literally anything to them with impunity. Five out of the 65 patients who underwent a lobotomy in 1953 died.[3] Fewer than a third 'recovered' sufficiently to be discharged or remain in Ypsilanti with reduced supervision. When the virologist Jonas Salk, who would go on to develop the polio vaccine, asked if he could test flu vaccines on patients at Ypsilanti, he was given carte blanche. They were even used as guinea pigs to test the effects of LSD. There was nothing and no one to protect them.

In the autumn of 1958, Yoder received a letter from Milton Rokeach, a renowned social psychologist, about an experiment he had devised. The idea was to use conversations as a tool to dispel the delusions of schizophrenic patients. Yoder must have felt that it at least couldn't hurt, and he told Rokeach he was welcome to try.

There was nothing to indicate that, for a time, Ypsilanti would be transformed into what with hindsight could fairly be described as a crime scene.

Cure through confrontation

Milton Rokeach had been born in 1918 as Mendel Rokicz in the Polish village of Hrubieszów. His father, a rabbi in a community of Hasidic Jews, had emigrated to America and then, when Mendel was seven, had his family join him. In the big Jewish community in Brooklyn, Mendel received a strict Orthodox education.

A few months before his death in 1988, Rokeach, writing to express thanks for a prestigious award, looked back on his youth. He had felt torn between his Hasidic Jewish identity – with a family who expected him 'to continue in the footsteps and traditions of my orthodox Jewish forefathers' and become a rabbi – and his growing left-wing sympathies, nourished by the Marxism of his friends.[4] But in Marxism, too, he had been struck by dogmatism, rigidity and orthodoxy. The similarity between ideological and religious dogmatism, he explained, would become the focus of his later work in social psychology.

Studying at Brooklyn College, he was part of an ambitious community of second-generation Jewish immigrants, taught by young and equally ambitious Jewish professors such as Solomon Asch and Abraham Maslow. He was still being tugged in opposite directions, but one thing was clear: he would not become a rabbi.

Shortly after Rokeach started his doctoral research at the University of California, the bombing of Pearl Harbor dragged the United States into the Second World War. Rokeach enlisted,

Milton Rokeach, 1974.

joining a team of psychologists who tested the professional aptitude of air-force cadets. After the war, back in Berkeley, he obtained his doctorate with a dissertation on the link between social prejudice and mental rigidity. In 1947 he accepted a post at Michigan State University, where he remained for 23 years.

A year before Rokeach sent his letter to Ypsilanti, the psychologist Leon Festinger had introduced the concept of 'cognitive dissonance'.[5] Festinger was a son of Russian-Jewish immigrants and, just like Rokeach, had grown up in Brooklyn. Both men sought to tackle concrete social problems through socio-psychological insights. Cognitive dissonance is based on the premise that individuals try to harmonize their beliefs, values and behaviour as much as possible. This 'consonance' is threatened when discrepancies arise between what someone believes, says and does. To give one of Festinger's examples: if you smoke, while at the same time believing that smoking is bad for you, the result is cognitive dissonance. You can restore consonance by giving up smoking. But you could also go in search of evidence that smoking isn't really so harmful or try to convince yourself that the advantages outweigh the disadvantages. What Rokeach was planning to do in Ypsilanti in many ways resembled a large-scale study of cognitive dissonance focused on schizophrenic patients and their delusional systems. They were the ideal guinea pigs. They were open about their delusions, had all the time in the world, were easy to get at – and had no means of escape.

What Rokeach was envisaging did have historical parallels.[6] In a commentary on a book of essays on crime and punishment, Voltaire relates the tragical history of Simon Morin. In 1663, inspired by a vision, Morin went about Paris publicly announcing that he was Jesus Christ. A local magistrate ordered that he be confined in a madhouse. By coincidence, another new patient at the madhouse claimed that he was God the Father. Morin found this utterly ridiculous, so much so that he realized he himself had been suffering from a delusion. He managed to convince the authorities that he was sane again and was discharged. Once at liberty, though, he relapsed, and he was soon telling whoever would listen that he was Christ. Worried that he might found a sect, the authorities condemned him to death – just like the real Christ. He died shortly afterwards, not on the cross, but by being burnt at the stake.

Exactly what Rokeach intended by retelling this story about Simon Morin remains a bit unclear. He cited it as an example of 'cure through confrontation', albeit that the cure was temporary.

He followed it up with a reference to a more recent case that had been published in 1958.[7] This time it concerned two women who were both convinced that they were the Virgin Mary, and who had met by chance on a hospital lawn. After some polite back and forths about which of the two must be mistaken – 'Why you can't be, my dear' – one of them asked a psychiatrist (who had been eavesdropping) what Mary's mother had been called. When he answered 'I think it was Anne', the woman cheerily declared that if the other patient was Mary, then she must be Anne. The women gave each other a hug and parted ways. The one who had been so accommodating as to swap her delusion was soon well enough to be discharged.

Confronting the deluded with the similarly deluded did not, it seemed, have a great success rate. It had not effected a permanent cure in the case of Simon Morin and, in the case of the two Marys, it had left one patient unchanged and prompted an alternative delusion in the other. Yet Rokeach felt sufficiently emboldened to experiment along these lines himself. He dashed off a letter to the medical superintendents of five large psychiatric hospitals in Michigan, asking if any of their inmates suffered from delusional identity.

Clyde, Joseph and Leon

Although the hospitals collectively housed some 25,000 patients, the pickings were disappointingly slim. No Napoleons, Rokeach noted sadly. No Khrushchevs or Eisenhowers either. There was, however, a Cinderella, two patients who claimed to be members of the Ford family and a 'Mrs God'. The picture was better when it came to Christs: there were six of them. Three had to be disqualified, though, because they were clearly suffering from brain damage. Of the remaining three Christs, two were already residents of Ypsilanti, while the third was an inmate of Kalamazoo State Hospital, some 180 kilometres (112 mi.) away. He was swiftly transferred to Ypsilanti, where the three men were given beds next to each other, ate at the same table and were all employed in the laundry. The psychiatric staff of Ypsilanti had cooperated enthusiastically in the project, Rokeach wrote. 'All of [them] shared my hope that the research we were about to engage in might lead to results of considerable scientific importance and, furthermore, to significant improvements in the mental state of the three patients.' The three Christs were given fictional names in his report: Clyde Benson, Joseph Cassel and Leon Gabor. Each of them was seriously mentally ill.

Clyde Benson was the son of a carpenter – a striking detail, given his subsequent delusion. His parents worshipped at a small Protestant Fundamentalist church. At the age of 24 he married Shirley, the daughter of a wealthy farmer. The couple lived with his parents for the next ten years. When Clyde was in his early forties, a string of misfortunes struck. First, Shirley died after an abortion. Then, in quick succession, his mother, father and father-in-law died too.

At one fell stroke, Clyde thus lost every older relative on whom he'd relied for advice, even as a married man. He tried to persuade the oldest of his three daughters to put off her marriage and keep house for him and her two younger sisters, but she refused. By now he was well-off, having inherited a large amount of property, but he was also drinking heavily. In 1934 he got married again, to a woman who was herself a heavy drinker. Desperate to buy booze, he sold his land, then his stock, then his furniture. Seven years on he was bankrupt. His wife left him. He was now 53.

On being arrested for drunkenness, he became violent. After tearing up his prison bedding, he ripped off his clothes and stood naked in front of his cell window, raving that he was Christ and could hear Shirley's voice coming from a plane. He was committed to Ypsilanti. The diagnosis was paranoid schizophrenia.

*

Joseph Cassel had grown up in a French-Canadian community in Québec. He was a bright child who read a great deal, especially English literature. His father – who would not allow English to be spoken in the home – disapproved of his passion for books and soon took him out of school. When Joseph was sixteen, his mother died, and he was taken in by his deeply religious grandmother. As a young man, he emigrated to Detroit, hoping to become a writer. At the age of 24, he married a young woman called Beatrice.

Right from the start of the marriage, she told Rokeach, Joseph had been cold and undemonstrative. Kissing was unhealthy, he thought, and he didn't allow Beatrice to touch his face. Sex between the couple was surrounded with rituals. Joseph did not like socializing; instead he preferred to read. He said that he was writing a book himself.

When their second daughter was born, he asked suspiciously who the father was. Was it the neighbour? Beatrice was deeply hurt. When their third daughter arrived, and he again asked who the father was, she angrily replied 'Dr. Jones'.

In the autumn of 1938, Joseph abruptly quit his job as a clerk in a department store, insisting that his wife get a job so that he could write. They sold their house, moving in with relatives. He started to show signs of paranoia, claiming that people were poisoning his food and tobacco. Sometimes he made Beatrice swap teacups with him to make sure she was not poisoning him. During this time, he secretly bought huge quantities of books, plunging the family into poverty.

By March 1939 the situation had become untenable. Joseph was committed to a mental hospital. Beatrice placed their three little girls in Catholic institutions. She took her husband's manuscripts, which filled two large boxes, and threw them away unread.

The religious mania didn't happen until much later. In 1949 – by now Joseph was in Ypsilanti – he wrote to one of his daughters that he was God and ruler of all. When one of the hospital's psychiatrists asked if he had ever seen God, he answered: 'I can't very well see God when I *am* God.' He, too, was diagnosed with paranoid schizophrenia.

※

Leon Gabor had been born in 1921 to a Hungarian-American family. His father had left when Leon and his little brother were still small. Their mother, Mary, who hardly spoke any English, raised them single-handedly, scraping by on her earnings as a cleaner. When she went out to work, she would lock the little boys in the bedroom. Her life revolved around work and the church. She never missed Mass, and spent much of her time between services praying, spurred on in her zeal by the voices she heard in her head. In her spare hours she lavished more care on her garden than on her children. Her priest told Rokeach in broken English that he thought her fanaticism unhealthy: 'Leon, every day worse. She not cooking for boy, crackers and tea, not food for a boy growing. She not cleaning her house, praying, praying, all the time praying.'

Despite all this, Leon was an above-average student at the parochial school. He went on to attend a pre-seminary school for a few years, but for some reason he was expelled at the age of seventeen. He eventually found a job as an electrician. His wages were given to his mother, who had now stopped working and divided her time between her garden and church.

In 1942 Leon joined the U.S. Army. Assigned to the Signal Corps, where he worked on radar reconnaissance, he was repeatedly decorated for heroism in combat. After his honourable discharge in 1945, however, things began to go downhill. He started various courses of training, only to abandon them. He complained of chronic exhaustion and was eventually fired from his job because of absenteeism. Even though he was now nearly thirty years old, his mother kept a tight hand on the purse strings. She also wouldn't allow him to buy a radio: it would keep her from hearing the voices in her head.

In 1954 Leon began to hear voices too. They told him he was Jesus. One day he locked himself in the toilet and refused to come out. Mary fetched the priest, the priest fetched the doctor and Leon was briefly admitted to a mental hospital. After being discharged, he moved back in with his mother.

Leon must have had a clear idea of what Jesus looked like: his mother's walls were covered from skirting board to ceiling with prints of Jesus, the Virgin Mary, prayer cards, crucifixes and pictures of saints. One day he went berserk, ripping all these religious relics off the wall and smashing them. His mother happened to come back from church in the middle of this rampage. When she tried to stop him, he threatened to strangle her. No more false images, he yelled. From now on, she must worship the real Jesus: him.

The result: Leon was marched off to a mental hospital. The diagnosis: paranoid schizophrenia.

Once Leon had been transferred to Ypsilanti, Rokeach had all his test material in one place. His protocols for the conversations were ready; his assistants had been instructed. The experiment could begin.

'You have to be very careful what you say'

On 1 July 1959, Clyde, Joseph and Leon were invited to come to a small room off Ward D-23, where Rokeach and his assistants were waiting for them. There was a tape recorder on the table.

Rokeach introduced himself and asked Joseph to do the same. At the time, Joseph was 58 and had been hospitalized for two decades. Half his front teeth were missing. He wore three pairs of socks in layers: yellow, pink and yellow. His pockets bulged with books, magazines, pens and tins of tobacco, as well as large white rags that he used as handkerchiefs. 'My name is Joseph Cassel,' he said. Rokeach asked whether he had anything to add to that. 'Yes, I'm God.'

After that it was Clyde's turn. Aged seventy, he had by now spent seventeen years shuffling about the ward with his Christ delusion. All but toothless, he spoke indistinctly and was very hard to understand. His name was Clyde Benson, he said, but went on to add that he had other names and that he had 'made God five and Jesus six'. When Rokeach asked whether that meant he was God, he answered, 'I made God, yes. I made it seventy years old a year ago. Hell! I passed seventy years old.'

Leon's introduction was more elaborate. He must have taken close note of the prayer cards at home. Dressed in a white jacket and white trousers, the 38-year-old would stride with silent dignity through the ward, often holding his hands in front of him, palms up. Tall, lean and erect, he had the face of an ascetic.

'Sir,' Leon began, 'it so happens that my birth certificate says that I am Dr. *Domino Dominorum et Rex Rexarum, Simplis Christianus Pueris Mentalis Doktor* [Lord of Lords, King of Kings, Simple Christian Boy Psychiatrist].' He graciously permitted the others to address him as Rex. According to him, his birth certificate stated that he was the reincarnation of Jesus Christ of Nazareth. He attributed the fact that he was in Ypsilanti to electronic manipulation, regarding himself as the victim of a plot to turn him into a robot.

At this point he was interrupted by Rokeach, asking if Joseph wanted to say something. 'He says he is the reincarnation of Jesus Christ,' Joseph answered. 'I can't get it. I know who I am. I'm God, Christ, the Holy Ghost, and if I wasn't, by gosh, I wouldn't lay claim to anything of the sort. I'm Christ. I don't want to say I'm Christ, God, the Holy Ghost, Spirit. I know this is an insane house and you have to be very careful.'

Here Leon interrupted, telling him not to generalize. It was wrong to call everyone in Ypsilanti insane, he protested, because there were people there who weren't insane. 'This is an insane hospital, nevertheless,' Joseph insisted.

Rokeach asked Clyde what he thought. He, too, stood his ground, claiming that he represented the resurrection and was God, Christ and the Holy Spirit. This time it was Joseph who butted in, bemused as to why 'the old man' was saying that. Shortly afterwards, the two locked horns, both yelling that they were God and the other was lying. Rokeach had to step in and calm things down.

Leon had listened attentively and in silence to the two men arguing. At the end he protested against the meeting on the grounds that it was 'mental torture'. He announced that he would not be coming to them any more. But the next day, he joined in quite willingly.

In the second meeting, Rokeach enquired about the details of their delusions. He reminded Joseph that the previous day he'd said that he was God, Christ and the Holy Ghost. So that must have meant that he had created the world? 'That's correct,' Joseph confirmed. Rokeach turned to Clyde. 'Clyde, did you make the world, too?' But he refused to be drawn, saying he had better things to do than play games with patients. Rokeach drew Leon into the discussion, and he explained that there were two types of God: God Almighty who was spirit without a beginning and without an end, and creatures who were 'instrumental gods'. Some of this category of lower-case gods were hollowed out, while others weren't. He himself was a reincarnation of Jesus Christ of Nazareth, but he respected Joseph and Clyde as instrumental gods. To his irritation, he was frequently interrupted by unintelligible mumblings from Clyde. When it happened again, Leon couldn't contain himself: 'Jesus Christ! Let me get a word in, will you, please?' The rest of the meeting disintegrated into a confused row about who was what exactly. Clyde and Joseph rejected the status of hollowed-out instrumental gods. 'I'm not hollowed out. Not hollowed out at all! Clyde protested. Joseph again repeated that he was God, Christ and the Holy Ghost. Leon said that anyone who doubted him should go to Dr Yoder's office. There they could see for themselves that his birth certificate stated he was baptized as Jesus Christ. As far as he was concerned, that settled the matter. 'Sir, if you will excuse me, I do not care to sit in on any more discussions.' And off he went.

Machines

In the weeks that followed, the meetings took on a certain routine. Rokeach and his assistants arranged their test subjects around a table and would start the conversation. They often ended acrimoniously. On one occasion things got so heated that Clyde got up and hit Leon hard on the cheek. Leon didn't defend himself, just sat there serenely. Rokeach and his assistants had to pull Clyde off him. At the end of each meeting, they all stood to sing the patriotic anthem 'America the Beautiful'. At Rokeach's request, the three men took it in turns to chair the meeting, with Joseph carefully keeping track of whose turn it was. The chairman's tasks included handing out the cigarettes that the Christs received as a reward for their participation. As a result of all these meetings, a certain familiarity sprang up between the men, and there was less focus on who was the true Christ. Instead, they discussed a wide range of topics.

Joseph was the most erudite of the three. He regarded himself as a great writer. When the conversation touched on *Madame Bovary*, Rokeach asked him who the author was. Flaubert, Joseph told him. And when was it written? Around 1874. When questioned, he could give an accurate account of what the novel was about. For ten minutes or so, Rokeach wrote, it was as if they were having a perfectly rational conversation. But at the end, Joseph wove Flaubert and his novel into his delusion. 'You know, I really wrote *Madame Bovary*.' 'You did? I thought you just got through saying that Flaubert wrote it?' 'No, I did. Flaubert stole it from me. He took it to France.'

One day they were discussing sex. Clyde said that it had been eighteen years and ten months since he'd had a woman. But he couldn't have sex while he was on duty, as he put it. 'When this project is finished,' he announced, 'I can have all the women I want.' Joseph said he didn't miss women because he was too busy with his work, and he wouldn't have sex with a woman unless he was married to her. He wouldn't make any advances; it would be taking advantage of his power. But he went rather quiet, and when Rokeach asked why, it turned out the subject brought back nostalgic memories: 'I am thinking about the old days outside. On Saturday nights you take your girl to the movie theatre and feel her tits.' 'Sir,' exclaimed Leon, 'your language, please!' He went on to say that he'd never had sexual intercourse, and he looked forward to getting together with his wife. 'I have good news. Last night I thought about my wife and I got a hard-on, and I came without trimming the candle.'

After many more confrontations, each of the Christs had developed their own narrative about how the others could also believe that they were Christ. Leon stuck to the notion that the other two were instrumental gods. Clyde believed that Joseph and Leon were in fact dead. It was the machines in them that were talking. If you took the machines out of them, they would stop talking. When Rokeach asked where this machine was located, Clyde replied by pointing at the right side of Joseph's stomach. Pretending to go along with him, Rokeach asked Joseph if he would mind unbuttoning his shirt, so that Clyde could show where the machine was. Clyde started to feel around Joseph's stomach but couldn't find anything. 'That's funny,' he said, 'it isn't there. It must have slipped down where you can't feel it.' His faith in the machine inside the dead Joseph remained unshaken. For his part, Joseph said that as Clyde and Leon were patients in a mental hospital, they were clearly insane.

Six months or so after the discussions started, Rokeach and Leon came into sharp conflict. Rokeach had given a lecture to the Psychology Club of his university on the subject of his experiences with the three Christs up to then. The *Ypsilanti Daily Press* reported the event: 'Three mental patients – each claiming to be Jesus Christ – have been brought together at the Ypsilanti State Hospital. "The purpose of the experiment is to see what happens when a person's belief in his identity is challenged by someone claiming the same identity," says Dr. Milton Rokeach.' The article went on to report the provisional findings of the project: according to Rokeach, one of the test subjects had changed his belief about being Christ and had taken on another false identity. The other two, who had been hospitalized longer, still believed that they were Christ. One claimed that the other two were dead and were operated by machines inside their bodies. The other thought that the other two were crazy, and that it was not for nothing that they were in a mental hospital.

Rokeach had brought the newspaper clipping with him and gave it to Joseph, asking him to read it aloud. Claiming that his eyesight wasn't good, he passed it to Leon, who read the first few sentences to himself. His reincarnation was being ridiculed, he told the group, and the psychology was warped. Then he read the whole article aloud. Afterwards he tried to control himself, but he was visibly upset. He reproached Rokeach bitterly: the article didn't say what was written on his birth certificate, he was still Jesus Christ, he hadn't changed his personality at all. 'When psychology is used to agitate, it's not sound psychology anymore. You're not helping the person. You're agitating. When you agitate you belittle your intelligence.' At that point he got up and left, saying that he had to go to the toilet.

Rokeach simply continued with the remaining Christs, asking them who was being talked about in the article. Joseph explained that it was talking about three mental patients claiming to be Jesus Christ. Rokeach asked him if he knew who they were. Joseph said no. Did he really not have any idea? No, Joseph said, their names weren't in the article. Rokeach persisted. Joseph replied that someone who thought he was God 'should be sent to a hospital – not to be gotten out, not to be dismissed until he has gotten well'. Rokeach asked how you would know when he was well. 'When he claims he is not Jesus Christ anymore.'

When Leon returned from the toilet, he was still angry. He shouted at Rokeach that he was responsible for the information in the clipping and that he (Rokeach) had belittled ('deplored') his own intelligence. A person who was supposed to be a doctor or a professor should be lifting people up, guiding and inspiring them. Leon:

'I believe I could give a better lecture than some people who went to college twelve years.'
– *It's quite possible.*–
'I know I can with the help of the good Lord.'
– *You seem very angry. –*
'I'm angry at the evil ideal, not at you people. I feel sorry for you.'
– *It's human to feel angry. –*
'I'm angry at the evil ideal that has made a foolish-sound-ing person out of you. I sensed it in the first meeting – deploring!'
– *Deploring? Do you know I've come seventy-five miles in snow and storm to see you! –*

'It is obvious that you did, sir, but the point still remains, what was your intention when you came here, sir?'

– *What was my intention, if not to help you!* –

'I don't think so!'

– *I can get sore too!* –

'It's your privilege if you want to get angry at my speaking the truth.'

Rokeach managed to calm things down. The meeting ended as usual, with everyone rising to sing 'America'.

Weeks later, Joseph spontaneously brought up the subject of the clipping. 'There is an article,' he said, 'about three fellows being under observation at Ypsilanti State Hospital. One is God, one is Jesus Christ, and one is Napoleon or something. Hell! That makes you kind of scared. You wish you were somewhere else. What they ought to do is dissolve the meetings, not have any meetings, don't you think?'

After the experiment had been running for about half a year, something happened that brought the three men closer together. The day after Christmas 1959 Joseph complained of a pain in his stomach. After trying to treat it by rubbing his abdomen with a bleaching compound he'd found in a cupboard in the laundry, he was rushed to the infirmary. On New Year's Eve, Rokeach suggested that they hold the meeting at Joseph's bedside, and Clyde and Leon agreed. On the way there, they passed through a lobby filled with floral displays donated by local funeral parlours. They selected a bouquet for Joseph. When the little group arrived at Joseph's bed, Leon placed the flowers on Joseph's night table, carefully arranging them to the best effect. Joseph was immensely touched, commenting over and over again how nice the flowers looked and repeatedly thanking Clyde and Leon. The men held a meeting, ending with the usual rendition of 'America'. Leon begged some tobacco paper from the nursing aides and filled Joseph's empty tobacco pouch from his own. At the end of the meeting, everyone wished each other Happy New Year and shook hands.

Outside the meetings, the men often sought each other out. Even when a meeting had involved heated exchanges, with a lot of name-calling and the occasional wild punch, they would afterwards walk together to the recreation room where they pushed their chairs together. There, they would silently puff away their state allowance of tobacco, lend each other tobacco papers or light each other's cigarettes. When Rokeach wasn't there, they knew how to avoid acrimony. When they were by themselves, they would never raise the subject of who was the real Christ, or they'd merely respond with a mild 'That's your belief, sir.' A kind of camaraderie sprang up, unusual among schizophrenic patients, who tend to withdraw into their own private worlds.

But meanwhile, a year had gone by and each Christ was still comfortably clinging to his own delusion. However pleasant for the men themselves, this was not what Rokeach had intended when he began his experiment.

Tears

Clyde was more or less a lost cause. Years of alcoholism had left their mark, and the conversations largely passed him by. By contrast, Rokeach had his hands full with Leon. If the scenes with the three Christs are reminiscent of Ken Kesey's novel *One Flew Over the Cuckoo's Nest*, published two years before Rokeach's report, Leon resembles the rebellious Randle McMurphy, as played by Jack Nicholson in the 1962 film version. Leon was the only one who realized what Rokeach was trying to do: forcing them to relinquish their beliefs by making them clash with one another. It's not for nothing that it was he who was so indignant when Rokeach said he had driven 120 kilometres (75 mi.) through the storm and snow to 'help' them – helping was never the aim of the experiment, at best a possible consequence. Leon was irritated by the constant trouble-stirring and manipulation. He suggested that Rokeach, 'with that Jewish nose', might be the reincarnation of Caiaphas the high priest presiding over the trial in which Jesus was condemned to death. Just as McMurphy resists the authority of Nurse Ratched, Leon did what he could to sabotage Rokeach's attempts to make the men turn on each other. Now that, despite Rokeach's efforts, all three had retained their delusions intact, Leon was the logical choice when it came to testing a harsher form of manipulation.

Shortly before, Leon had changed his name of Rex to Dr R. I. Dung. The name was inspired by the biblical parable of the fig tree in Luke 13:6–9. For three years, the fig tree had produced no fruit, and the owner of the vineyard wanted to cut it down. The gardener suggested an alternative: he would fertilize it with dung, and if it still didn't bear fruit, then it could be cut down. For Leon, therefore, dung had positive associations. It was one of many occasions on which he explained changes to his delusions with references to Bible stories.[8] This change of name didn't go smoothly. Leon categorically refused to respond if he wasn't addressed as Dung. But the head nurse, Mrs Parker, whom Rokeach had invited to sit in with a meeting of the Christs, could not bring herself to do so: she disapproved of the name and found it embarrassing. Leon, always impeccably polite, explained that the name was really in the Bible – he could show her exactly where – but eventually gave her permission to address him as 'R. I.' The other Christs were vastly amused by his new name, but Leon stuck to his guns. He had a new delusion, too: that he had a wife called Madame Yeti – Leon had read an article about the Himalayan Abominable Snowman in a magazine. What would happen, Rokeach wondered, if Leon were to receive a letter from a character in his delusional system?

One day in the summer of 1960, a nursing aide handed Leon a letter. He said that a lady had approached him and asked him to deliver it to Dr Dung in Ward D-16. Leon thanked him politely and read it:

> My dear husband,
> I have been aware on Channel 1 that you have been wait-
> ing for me to visit you a very long time. If the good Lord
> permits I will visit you at the Ypsilanti State Hospital,
> Ward D-16 on this Thursday at 1 o'clock.
>
> > Sincerely,
> > Madame Dr. R. I. Dung.

Leon told the aide that his wife didn't write this letter, and he didn't mention it to the others. But on that Thursday, he left the ward shortly before the appointed hour – possibly a sign that Leon truly believed in the existence of his wife.

In a subsequent letter, she enclosed a dollar bill. Leon didn't care about money. He never spent the small veteran's pension he received: over the years, it had grown to the considerable sum of almost $1,000. He didn't think he was entitled to the money: after all, the account was in the name of Leon Gabor, and that wasn't him. But the sight of that dollar bill affected him, and he sat gazing at it intently. Then Rokeach noticed that he was fighting back tears – in vain. A teardrop welled up in each eye and trickled slowly down his cheek. Leon caught them on his fingers, scooping them into his mouth. Rokeach asked what he was doing.

'Tears are the best antiseptic there is,' Leon said. 'There's no use wasting tears.'
He began to examine the dollar bill, turning it over from one side to the other.
'I haven't seen one of these for years. I mean, to handle.'
He read the name of the Treasurer of the United States and the serial number.
– *Does the letter make you happy or sad?* –
'I feel somewhat glad.'
– *Is there something the matter with your eyes?* –
'Oh, they're smarting, sir, so I'm enjoying some disinfectant, sir – the best in the world: tears.'
– *Are you crying?* –
'No, my eyes are smarting because of some condition.'
– *You say you feel somewhat happy?* –
'Yes, sir, it's a pleasant feeling to have someone think of you.'

Rokeach then tested whether Leon was prepared to act on suggestions from his wife. Apparently omniscient and omnipresent, she wrote that she was beginning to find the meetings of the Three Christs a little dull. Did they always have to sing 'America'? Couldn't they mix things up a bit? What about singing 'Onward,

Christian Soldiers' instead? The next time Leon was chairman, he proposed that they sing the first verse of 'Onward, Christian Soldiers'. Afterwards, he told the group that his wife had suggested it. She continued to send him money, instructing him to buy a ballpoint pen with it and to give the change to Clyde and Joseph, which he did. Leon never wrote back – nor could he, since there was no sender's name or address on the envelopes. After receiving a few more letters, Leon asked the aide whether there was a woman in the hospital who called herself a female God, and whether – if she gave him another letter – he could ask her name and ward number. It was the beginning of the end of the letter episode.

When another letter arrived, Leon felt the coins inside and handed it back to the aide, refusing to accept it. When the aide made a second attempt to deliver it that evening, Leon again refused, saying that he wanted nothing to do with the money. He then wrote a letter himself, not to his wife but to the female psychiatrist in charge of Ward D-16:

> Respected Dr Broadhurst,
> Please return these three dimes to Madam Yeti Woman,
> I know you know who she is. Tell her I do not want any
> more donations, or letters. Tell her I trust in the sanity of
> God, the Ten Commandments of God.
>
> > Respectfully,
> > Dr. R. I. Dung.

After that, he categorically refused to receive any letters. Later it emerged that he thought they'd been written by Dr Broadhurst. He never knew who the author really was.

How could things have so suddenly gone awry? Rokeach and assistants concluded that they had underestimated his moral objections to money. The sums involved were small, and he could share them with Clyde and Joseph, but apparently it was the principle that bothered Leon. His wife had made him do things that were at odds with his conscience. He could only resolve the dissonance by radically changing his delusional system: from then on, his wife was no more. Leon could resume life as a Christ who would not be sullied by earthly mire.

Rokeach was forced to admit that he hadn't made much progress in his treatment of Leon. But there was still Joseph.

'Dear Dad'

The superintendent of Ypsilanti, Dr Yoder, was respected by his fellow medics (he was, among other things, President of the Michigan Society of Neurology and Psychiatry), but his patients truly worshipped him. His manner with them was fatherly; some addressed him as 'Dad'. He regularly received letters from inmates requesting his photograph. Of the three Christs, Joseph felt the strongest affection for Yoder. He'd gone to Yoder behind Rokeach's back, asking to talk to him 'man to man' about his impotence. Couldn't he be helped in some way? 'A man must have a hard-on. He feels better all around.' Yoder hadn't been able to help him. But it did give Rokeach an idea.

In the language of cognitive dissonance theory, Dr Yoder was a 'reference person' for Joseph, an authority figure who could influence his beliefs and actions. Might their bond of trust, Rokeach speculated, be used for a new manipulative experiment? From July 1960, Joseph received a series of letters – supposedly from Dr Yoder but actually written by Rokeach – addressed to 'My dear

Joseph'. Honoured by so much personal attention, Joseph usually opened them at the meetings with the other Christs and proudly read them aloud. As soon as he could, he would write back to 'Dear Dad'.

Rokeach had of course asked Yoder for permission before embarking on this venture. Yoder had consented on condition that the letters were always positive and encouraging. Rokeach interpreted this loosely, to say the least. After all, the experiment required two convictions to be made to clash – and now and then that involved heavy-handedness.

The conviction that he was Christ wasn't Joseph's only delusion. He also believed that he was called John Michael Ernahue and came from England (he regarded Ypsilanti as an 'English stronghold'). He also claimed that he was the governor of Illinois. This dual identity of God and governor caused some surprise in the group, but Joseph had an explanation ready: 'I have to earn my living, you know.' When he wrote to Yoder requesting a transfer to a mental hospital in his native country of England, Rokeach seized the opportunity to reply in Yoder's name, stating that this would unfortunately not be possible. According to the hospital records, he had been born in Canada and was a naturalized citizen of the United States. This answer disappointed Joseph but seemed to work. When, a few months later, Rokeach asked him casually where he had been born, he answered 'in Québec'. He also stated that he'd never been to England. Later, however, he relapsed into his England delusion, though he smilingly evaded further questions on the subject.

In a subsequent letter, 'Yoder' denied that Ypsilanti was an English stronghold and appealed to Joseph to recognize that. Joseph's response was to tear up the letter angrily. 'Yoder' also repeatedly urged Joseph to attend church, and to persuade Clyde and Leon to go too. 'I would suggest that Dr. O. R. Yoder mind his own business,' Leon responded tersely. Clyde was equally disinclined: 'I know more about church than they can talk about. I *am* the church. I'm saved.' Joseph defended the superintendent: 'He, Yoder, was just suggesting that I invite you two fellows.' Rokeach felt that it was time to step up the pressure on Joseph's delusions.

Joseph received a letter in which 'Yoder' announced that he was going to give Joseph a new miracle drug that could restore his sanity, courage and self-confidence and eliminate fear and anxiety. This drug – called *potent-valuemiocene* – was not yet generally available. But out of all the patients in Ypsilanti, 'Yoder' had decided to give it to Joseph, whom he loved 'like a son'. He had instructed the aides to give him two of these small but extremely powerful capsules each morning. The next day 'Yoder' received a grateful letter: 'This is swell – I sure need this drug.' A week later, 'Yoder' wrote to say that Joseph must be feeling much better now thanks to the pills and that he must now see Clyde and Leon for what they were: mental patients in a mental hospital.

On 4 May 1961, the screws were turned more tightly. That morning, Joseph went to the ward doctor as usual to get his daily pills but was told that he would no longer be receiving the drug: Rokeach wanted to see what effect this would have on him. Joseph immediately penned a desperate appeal to 'Yoder': the medicine had been so effective, curing the neuralgia in his cheek and shoulder – Yoder had himself prescribed it – why wasn't he to receive it any more? In a PS he wrote: 'I will continue to work as God, for the hospital, the English, the civilizations . . . the people of this world – I work for all!' To his relief, he was given his pills again the next day, but he received another letter from 'Yoder', this time with a fatherly warning. Working for the English? Did he still think that Ypsilanti was an English stronghold?

> I think we have had a discussion about this on an earlier occasion. This is a mental hospital supported by the tax-payers of the State of Michigan, and when you say 'I am getting better mentally and physically' I suppose that you mean that you are now more realistic about the fact that it is an American hospital and not an English stronghold.

In his letter thanking Yoder for the resumption of the medicine, Joseph wrote that he was well aware that Ypsilanti was a mental hospital and was gratified that it was supported by tax-payers, but he persisted in claiming that the hospital was an English stronghold, even though it was in the United States. 'Yoder' responded that he was disappointed to hear this; the pills were supposed to correct this erroneous belief. Since the pills apparently weren't working, he was seriously considering withdrawing the medication.

Joseph immediately dashed off three letters to avert this calamity. In a meeting with the other Christs, he tried to find a diplomatic solution: couldn't he just call it an American hospital, without explicitly denying that it was an English stronghold? But 'Yoder' harshly cut off that avenue. Since the medicine was clearly not doing him any good, he wrote, it must be stopped. The only way to prove that it was working was for Joseph to sign the following statement:

I, Joseph Cassel, do hereby state that
Ypsilanti State Hospital is not now and
never has been an English stronghold.

It was entirely up to Joseph whether he signed or not, 'Yoder'
wrote. He would only take a final decision when he had heard
from Joseph. He assured him that there was absolutely nothing
wrong between them, and that he would always love Joseph like
a father loved his son.

Joseph interpreted this to mean that he was not obliged to
sign, and he wrote to 'Yoder' that same day to tell him so. He hoped
that he could nevertheless keep his medicine. 'Yoder' quickly
quashed that hope, informing Joseph that the medication would
stop the next Monday unless he signed the statement.

After that, things moved quickly. Joseph read 'Yoder's' letter
aloud to the other Christs, unable to believe that the medicine
that was doing him so much good would be discontinued. There
must be some mistake, surely? He appealed to 'Yoder' in desperation: 'Please let me take this medicine, dad? As I am not guilty
of anything towards you, why, then, can I not take this medicine?'
The next Monday came the answer he had feared. In a brief letter,
'Yoder' informed him that the medication would be discontinued
with immediate effect.

Only then did Joseph give up – not the delusion, but his
attempts to keep the medicine. A disappointed letter was penned
to 'My dear Dad':

I was hopeful that I would partake of the medicine
(potent-valuemiocene) for a longer time, but since you
have decided otherwise I simply have to do without the

medicine. As you are the head doctor of this hospital, your saying for the discontinuance of the medicine is so valid that I am now helpless to do anything; in other words, I have to do without.

Rokeach had finally found out what he wanted to know. Not only was Joseph not prepared to relinquish his delusions, but his refusal to sign showed he didn't even want to *say* he had relinquished them. Rokeach had hoped that 'without excessive pressure' – those were his exact words – Joseph could be persuaded to sign the statement, creating dissonance between the delusion that Ypsilanti was an English stronghold and his public statement that it wasn't. The tension between what he believed and what he said might have compelled him to relinquish his delusion. But even at the prospect of losing medication he felt so dependent on and – worse still – of disappointing his 'Dad', he preferred to cling to his delusion. 'It is my opinion', Rokeach concluded, 'that even explicit threats or punishment would not have induced Joseph to sign the "oath".'

Even explicit threats and punishment?

What did Rokeach think he had been doing in the letters from 'Yoder'? A process that had begun a year earlier with flattery and friendly requests from a father to his son, followed by mild admonishments, had descended into exactly that: a shameless barrage of threats and punishments, with the sad result that a patient still believed not only that he was incarcerated in an English stronghold, but that he had disappointed his father by not wanting to lie about it.

Departure

On 15 August 1961 the researchers said goodbye to Joseph, Clyde and Leon. Fresh developments could no longer be expected. Rokeach now had to complete his report.

In the final chapter, he presented his analysis of the thoughts and actions of the three Christs. Its slant was heavily psychoanalytical. In his study of paranoid schizophrenia – the 'Schreber Case' – Freud had concluded that paranoid delusions of grandeur were motivated by latent homosexuality. Rokeach interpreted this more loosely: he thought delusions of this type were prompted by confusion about sexual identity. Before being admitted to Ypsilanti, all three men had experienced troubled sexual relationships. According to Rokeach, the resultant confusion was reflected in their relationships with delusional characters. Leon, especially, had trouble regulating his sexual impulses to fit the relationship with his imagined wife. This Freudian approach to delusions of grandeur was jettisoned long ago – though intense, persistent confusion about sexual identity *is* seen as a risk factor for mental health issues.

Nevertheless, the study did highlight something that a paranoid patient refuses to relinquish under any circumstances. However bizarre and exotic a delusional system, it must always possess order and logic, both internally (between the various elements) and in relation to reality. If that order is threatened, either internally or externally, the patient scrambles to restore it. All three Christs rationalized the existence of others claiming the same identity. Sometimes their logic was clumsy and primitive. Clyde simply claimed that the other two were dead, while Joseph denied being one of the patients referred to in Rokeach's newspaper clipping. Leon's approach was more nuanced: he did not deny that there were two other men claiming to be Christ, but he wove their claims into a delusional world that featured not just the one true Christ – himself – but also lesser gods, 'gods with a small g'. In this way, order was restored.

The letters that Joseph and Leon received posed an internal threat to order as they saw it. Requests that didn't affect their delusions could easily be agreed to: fine, from now on they would sing 'Onward, Christian Soldiers'. But in his letters about the English stronghold, 'Yoder' required Joseph to do something quite impossible: relinquish the very delusion on which his new identity rested. And through the dollar bills sent by his 'wife', Leon was manoeuvred into a position he couldn't reconcile with the identity of Christ, who had driven the moneylenders from the Temple and wanted nothing to do with earthly wealth. Both men solved this conflict in the same radical way: they broke off contact. But the very fact that adjustments were necessary proved that they still kept a toe in reality and felt a need for consistency. By way of an experimental control, 'Yoder' had also occasionally sent letters containing money to Clyde and Joseph, but they had simply stuffed them in their pockets without further ado. Clearly, their self-image of Christ was not quite as austere as Leon's.

The need for logic and consistency was also expressed in the three men's attempts to interpret every event in terms of their delusional systems, from the appearance of a female psychiatrist to the arrival of an attractive female research assistant. Chance does not exist to those who suffer from paranoia: everything has significance. Rokeach had the impression that Leon, especially, spent the entire time mentally knitting together experiences to form a consistent whole.

What went on in the mind of the withdrawn Clyde was a mystery to everyone; perhaps not very much. But the delusions of Leon and Joseph were a work in progress. They required constant tuning, adjustment, revision. Rokeach seemed to be more focused on their origins than on the function that they had. The delusions had a purpose: they helped the three men to cope in what to them were chaotic surroundings. They furnished them with a clear role and a reasonably predictable relationship with the outside world. Their delusional identity was more important than anything else – it was really the only thing they had. So it was not for nothing that they defended it tooth and nail. Whereas Rokeach regarded the Christs' delusions as something he needed to 'free' them from, to the men themselves they were a place of refuge, a shield. The delusions enabled them to preserve their self-respect, and the more forcefully Rokeach tried to claw them open – as with the letters from Madam Yeti Woman and 'Dr Yoder' – the further the men retreated. 'The cave, the last stronghold, becomes more and more inaccessible to light,' Rokeach wrote in his epilogue. To him, 'the light' was the contact with reality that needed to be restored; to them, 'the cave' was a shelter where they were safe.

Almost as an afterthought, at the end of his book, Rokeach mentions that, by way of control group, he had simultaneously started a similar experiment with three female schizophrenic patients. One woman believed herself to be Cinderella, another that she was a member of the Morgan family (a puissantly wealthy banking dynasty) and the third that she was bewitched. They too were placed in a group, worked together in the laundry, slept side by side in the dormitory and were confronted with each other's delusions in daily meetings. But Rokeach could summarize the outcome of the experiment in a few sentences. They did not quarrel and, six months on, their behaviour and delusions remained completely unchanged. He had abandoned the experiment, mainly out of boredom. The lack of conflict apparently made his dealings with the three women highly tedious.

After Rokeach and his team left, the daily meetings stopped. Yet the three Christs continued to seek each other out, in combinations of two or three. It put Rokeach in mind of the cohesion shown by a conflict-ridden family. Its members aren't always on speaking terms, but they're reluctant to sever ties altogether.

Demolition

Dr Yoder retired in 1964, the year that Rokeach's book was published, after a career spanning 42 years in mental health care. The American Medical Association elected him Doctor of the Year. It is unclear whether, having given his consent, he knew about the letters that were written in his name. It may be that he first found out on reading Rokeach's book.

The Three Christs of Ypsilanti was mostly well-received. Helen Harris Perlman of the University of Chicago spoke for many reviewers when she wrote that the study read like a novel and was full of fascinating observations. She praised Rokeach for 'the warmth and tender humour toward his subjects'.[9] A year later, Barbara Nachmann wrote that *The Three Christs* had appealed to a lay public, fascinated by the excerpts from the conversations with the three schizophrenic men, but she was scathing about the shortcomings of the researchers.[10] Rokeach never again wrote about his work at the clinic. He remained active in social psychology research, inspired by left-wing politics. In 1971 he testified for the defence in the trial of a 27-year-old Black Panther accused of murdering a police officer.[11] He argued that selecting a jury from a list of registered voters increased the likelihood of negative bias towards a suspect who was poor, Black and also a member of the Black Panthers.

Like Dr Yoder, Rokeach received prestigious awards, including an honorary doctorate from the Sorbonne. In 1984 he was given the Kurt Lewin Memorial Award and, four years later, the Harold Lasswell Award. He died in 1988 at the age of 69.

At the time, attitudes to government care of the mentally ill were changing. Large institutions were to be replaced with small-scale care facilities. A start was made on this transition, but the results were woefully inadequate. Opportunistic arguments invoking anti-psychiatry and the danger of institutionalization, combined with an overly rosy estimation of the effects of psychotropic drugs, led to drastic cuts. Not long after, the government decided to stop financing large psychiatric hospitals. Aid was dismantled with dizzying speed. In 1991 the governor of Michigan ended state support for Ypsilanti at the stroke of a pen. The hospital closed that same year. The consequences for the patients were disastrous. Some had spent half their lives in Ypsilanti, had indeed become institutionalized and had scarcely any contact with their families – if they still had any. They were just dumped on the streets and often remained there, homeless, displaced. No longer patients, they were now dubbed 'people with mental health issues', with a high likelihood of getting into trouble with the police. Within a few years, Ypsilanti's empty buildings fell into ruin, just like other large mental health hospitals all over the United States. In 2006 the immense complex was demolished and the land sold to Toyota. Dr Yoder lived long enough to experience Ypsilanti's demise. He died the year it closed, aged 97.

Rokeach's experiment continues to fascinate to this day. In 2011 *The Three Christs of Ypsilanti* was reissued in the series 'New York Review of Book Classics'. In the introduction, the author Rick Moody rightly praises its literary quality, with Rokeach always able to find the right words. For that reason, he ranks it alongside other famous case studies, such as those of Sigmund Freud, Alexander Luria and Oliver Sacks. He hides his reservations about the moral side of the experiment in a compliment: 'There's an earnestness in Rokeach, both during and after the experiment – no matter his theoretical naïveté and ethical lapses. There's an earnestness in any attempt to reach a schizophrenic on her or his own terms.'[12]

It is telling that the dissident voice in the chorus of favourable reviews is that of Jenny Diski. Committed to a mental hospital at the age of sixteen, Diski was taken under the wing of the author Doris Lessing and later became a writer and literary critic. Rokeach had claimed that subjecting the delusions of psychotic patients to strain was acceptable because 'there would be little to lose and, hopefully, a great deal to gain.' Diski continued in scathing vein:

This is a very magisterial 'non-deluded' view of who in the world has or has not little to lose. Evidently, the mad, having no lives worth speaking of, might benefit from interference, but if they didn't, if indeed their lives were made worse, it hardly mattered, since such lives were already worthless non-lives. It also incorporated the bang-up-to-the-moment idea that if you want to know about normality you could do worse than watch and manipulate the mad. The three Christs themselves, however, were of the certain opinion that they had something valuable to lose and made truly heroic efforts, each in his own way, to resist, as well as to explain to Rokeach and his team that their lives had considerable meaning for them. All of them, though Leon in particular, had a very clear understanding of what it was to be deluded, why it might be a useful option to choose over normality, and who did and didn't have the right to interfere in their self-selected delusions. Over the course of the research, each man indicated how far he was prepared to go along with Rokeach, how much he valued what was on offer, and when his boundary had been reached. And they did it with more than ordinary grace and dignity.[13]

That 'little to lose and, hopefully, a great deal to gain' is repeated elsewhere in the book in more or less the same words. In his ethical justification of the manipulative letters – properly submitted in advance to the staff of Ypsilanti – Rokeach stated that they would start very carefully, monitoring Leon's and Joseph's reactions step by step, and of course would stop directly if they saw that the letters were causing distress. But at the moment when he should definitely have stopped, when both men were visibly upset, he actually increased the pressure. In the case of both patients, Rokeach breached his self-imposed ethical limit. Diski is right: the pleas of Joseph to his 'dad' must have been read through the perspective of someone who found that his despair fell in the category of 'little to lose'.

In 2017 a film was made of the experiment, directed by Jon Avnet, with virtually the same title as Rokeach's book. A melodramatic version of events, it features an absurd role reversal. The tender-hearted, paternal Dr Yoder is portrayed as 'Dr Orbus', an arrogant, manipulative and ambitious figure. When the first reports appear in the press of the intriguing experiment going on in Ypsilanti, he gets rid of Rokeach ('Dr Stone', played by Richard Gere) and takes over the meetings with the Christs, ultimately driving Joseph to suicide. By contrast, Rokeach is portrayed as an idealistic, thoroughly empathetic character who manages to obtain the trust of the Christs, sticks up for them and tries to prevent them from being subjected to electroshocks or other coercive forms of treatment.

In reality the picture was quite different, to say the least. In a podcast made in 2014, Dick Bonier and Ron Hoppe, two graduate students who had been on Rokeach's team from the first meeting with the Christs, stated that as the experiment progressed, they became increasingly unhappy.[14] Day after day, they followed the three men around, recording their behaviour – typically for eleven hours at a time. Rokeach, by contrast, mostly sat at his desk in the university, appearing only for the meetings. The two students became attached to the test subjects. Bonier grew fond of Clyde, who turned out to be a great raconteur, talking about his former life in occasional moments of clarity. Hoppe had long conversations with Joseph about the books he was reading (and which Joseph threw out of the window when he thought no one was looking). The students were increasingly distressed by the pressure Rokeach put on the three men, eventually summoning up the courage to confront him about his confrontational approach. He heard them out, said that he didn't agree – 'and that was that'. The experiment continued. When Bonier and Hoppe left to continue their studies, their place was taken by Mary Lou Anderson. In the podcast, they describe the final stage of the experiment with a mixture of outrage and pity.

According to them, Rokeach deliberately encouraged Anderson – an attractive young woman – to flirt with Leon. This chapter of his book indeed makes for painful reading. Leon soon lost his heart to Anderson. He would ask to speak to her alone after the meeting, and when she agreed he always tried to make the plenary meetings as short as possible, so that she had more time for him. He brought along lists of topics to be discussed and prolonged the conversations until he really couldn't think of anything else to say. She didn't come at weekends. During the Friday meetings he was sulky at the prospect of not seeing her the next day; on Mondays he was sulky because she hadn't been there the previous days. His infatuation placed a huge strain on him: he felt angry and guilty about his sexual feelings for her. When the tension became unbearable, he made a mask with a blindfold out of cardboard, cellophane and rubber bands so as to be shielded from her. He wore it for weeks, even during his work, so as not to betray that it was Anderson he didn't want to see. When she came back after a week's holiday, he refused to talk to her. Whereas previously, she'd been the only person from whom he would accept a light for his cigarette, she was now the only person from whom he *wouldn't* accept a light. Rokeach had hoped to turn Anderson into a 'positive reference person', so as to use her to get a grip on Leon's delusions, but this failed, just as all his other manipulations had failed. Leon once again became withdrawn. 'The truth is my friend,' he declared, 'I have no other friends.' Fifty-three years on, Bonier and Hoppe still found it difficult to talk about this episode.

In 1984 Rokeach wrote an afterword to a new edition of his book. Twenty years had passed since his last meeting with the Christs. None of the three men had been cured. In 1970 Clyde, by then in his eighties, had been discharged into the custody of his family. Joseph had died in Ypsilanti in 1976. Leon – who must then have been around sixty – was still there.

Rokeach had re-read his book with mixed feelings. Looking back, he realized that while he thought he had set up a confrontation between three patients, it had in reality been a confrontation between *four*. He had overlooked himself. He, too, had been suffering from a delusion – that he could cure the others of their delusions through his omnipotent and omniscient actions. In truth, four men had believed themselves to be Christ. He now saw the book as ending differently: 'While I had failed to cure the three Christs of their delusions, they had succeeded in curing me of mine – of my God-like delusion that I could change them.'

In the run-up to his experiment, Rokeach had scoured the annals of psychiatry in search of parallels. He had found Voltaire's account about Simon Morin and his Christ delusion. But Morin's case ended badly: the confrontational method had only worked briefly. Around 1800 Pinel recorded a similar case that Rokeach probably never knew about. Three of the patients in the Bicêtre who believed themselves to be Louis XVI (he had been guillotined in 1793) were arguing fiercely about which one of them was the real king. Tempers rose, and it looked as though they might resort to fisticuffs. Marguerite, the wife of the superintendent, Jean-Baptiste Pussin, had tactfully solved the conflict. She took one of the men aside and said, with great seriousness, 'How happens it that you should think of disputing with such fellows as those, who are evidently out of their minds: we all know well enough that your Majesty alone is Louis XVI.'[15] Flattered by this deference, the patient stalked off, giving the two others a disdainful look. The same ruse worked with the second Louis XVI, leaving the third in undisputed possession of his title and ending the quarrel.

In other words, Madame Pussin had done the opposite of what Rokeach did: she had respected the identity her patients had created, had played along with their delusion, had spared their feelings and thus saved them from further humiliating confrontations. Admittedly, none of the three kings had relinquished their delusion. But neither had the three Christs. And it would take the fourth twenty years to do so.

TWO

Living in the Knowledge of Being Dead

In 1686 Théophile Bonet, a third-generation physician from Geneva, had an amazing tale to tell.[1] The story took place in Copenhagen.

An elderly woman had a stroke that left her half paralysed and unable to speak. To everyone's relief, after four days her condition began to improve. But after recovering the power of speech, she gave her daughter and the maid a strange task. She wanted them to dress her in a shroud that she had lying ready in a cupboard and place her in a coffin – because, as she put it, she was 'already dead'.

Her daughter did everything she could to talk her out of this crazy notion, but the woman only flew into a rage, cursing at the cruelty of being denied this last service. She got into such a state that there was nothing for it but to wrap her in the shroud and lay her in the coffin. She even helped them clip her hair back so as to look as neat as possible, smoothed down the shroud and then fell asleep contentedly.

Of course, the daughter and the maid couldn't bear to leave her like that, so they stealthily put her in bed. But when she woke up, she was still convinced that she was dead and had to be buried. So it was back into the coffin.

The next morning, she got up. Daily life continued as normal; there was nothing wrong with her health. She had now developed an extra delusion, though: she believed that she and her daughter were in Norway, and she wanted to return to Copenhagen. Once again, no amount of persuasion could rid her of this idea. When she got impatient and it looked as though she might just set off by herself, someone came up with a ruse: they put her in a carriage, drove around the town and then came back home again. The trick worked: on return, she thought she was now back from Norway. But the delusion of being dead did not disappear. She spent the nights in the coffin. From time to time, she would invite other deceased people to come and visit. She would cook for her dead guests and chat to them.

Two years after Bonet published this account of a woman who thought she was dead, another case turned up in the Netherlands. In 1688 Jan Six, a wealthy and influential citizen of Amsterdam (his portrait was painted by Rembrandt), recorded a 'curious history' concerning the wife of his barber. The author Geert Mak paraphrased it in his biography of Six:

> She claimed 'that she was without a heart and soul, that she was dead but appeared to walk, that she was merely a corpse.' She tried to fast, and when she failed to do so she said that she had eaten a little 'only for her mouth'. To prove this to her husband she ate then rat poison and said, 'half laughing', 'See, do you now believe that I'm dead? Because even though I eat rat poison, it doesn't kill me, because I'm long dead.' 'And an hour after that she died. She was buried in 1688 on the first of June in the Nieuwe Kerk in Amsterdam.'[2]

A century later, the Mennonite pastor Cornelis Sorgdrager started to keep a journal of events on the Dutch island of Ameland. His first entry read: 'In January a maid of twenty died in holm [Hollum, the island's largest village]. Each day she fancied that she would die and would therefore take no nourishment, and continuing thus did at length perish.'[3] Sorgdrager would keep his journal for another 47 years, yet another such case did not recur.

A dozen or so instances of other people who believed themselves to be dead can be found in the notebooks of surgeons and the archives of lunatic asylums. The remarkable case of Lucas Kier, a tobacco farmer of Amersfoort, who in 1835 committed a murder while in the grip of such a delusion, will be discussed in the next chapter.

The dominant tone in these early accounts is one of astonishment and disbelief. Only as neurology and psychiatry began to develop into official disciplines in the 1880s and death delusions were recorded in professional case studies did they gain the status of a syndrome in their own right. In June 1880 the psychiatrist Jules Cotard presented to his colleagues at the Société Médico-Psychologique in Paris the case of 'Mademoiselle X.', a woman who claimed she no longer had a brain, nerves or intestines.[4] She believed that she consisted only of a hollow carcass of skin and 'the bones of her disorganised body'.[5] She had been sectioned in 1874 at the age of 43, after repeatedly trying to set fire to herself to be rid of the corpse. Mademoiselle X. suffered from depression. She thought she was damned because of sins committed around the time of her First Communion. Her past, she said, was a tissue of lies and crimes. She had gone to priests and doctors for help, but in vain. When Cotard examined her, it turned out that much of her body was insensitive to pain. He jabbed her repeatedly with needles – 'quite deeply', he assured his audience – without any reaction.[6] She was very consistent about living life as a dead person. She had stopped eating – because of course that was completely unnecessary – eventually dying of starvation. Indeed for many patients with this syndrome, the delusion ultimately becomes reality.

But how can you *think* that you are dead? After all, when you're dead, you can no longer think. Doesn't the delusion disprove itself? Can you really believe something so impossible? In the case of Mademoiselle X., the situation was even more paradoxical. She believed that she was dead but at the same time that she would live for ever. That delusion of immortality, Cotard explained, was weirdly logical: if you've died and yet continue to live, then apparently you *can't* die – so you must be immortal.

For more than a century, Cotard's syndrome remained a philosophical, psychiatric and neurological mystery. Only in the 1990s did a possible explanation emerge. It seemed that a different

Jules Cotard (1840–1889). This photograph, a detail from a group shot, dates from 1861, when Cotard worked under Jean-Martin Charcot in the Salpêtrière.

disorder – Capgras syndrome – might shed light on its origin. In recent years, various studies have used imaging technology to try to pinpoint which types or combinations of neurological damage might trigger or perpetuate Cotard's syndrome.

The *négateurs* of Jules Cotard

Jules Cotard was the son of a printer, but he opted for a different career path: that of *aliéniste*, or psychiatrist.[7] He studied under the renowned neurologist Jean-Martin Charcot in the Salpêtrière hospital in Paris. In his book *In Search of Lost Time*, Marcel Proust apparently based the character of 'Dr Cottard' – a skilled but shy and overly modest physician – on the introverted and rather gloomy Cotard.[8]

In 1874 Cotard moved to Vanves, where he took up a post in a *maison de santé*. These institutions were private clinics, usually in a rural location, where patients from wealthy families could be discreetly admitted and nursed under circumstances that were vastly better than in most of the big lunatic asylums in cities: individual rooms rather than overcrowded dormitories, gardens and parks rather than depressing courtyards. Set against this was the fact that every discharge meant a loss of income, which could lessen keenness to cure patients.

For years, Cotard worked on a treatise about medicine, philosophy and society. It was to be his magnum opus. Sadly, just before completion, 'catastrophe' struck (the details are unknown) and the manuscript was lost.[9] He could not bring himself to resume this work. In any event, he was not granted the time. As Shakespeare wrote, 'When sorrows come, they come not single spies but in battalions' – and Cotard's life was no exception. In August 1889 his oldest daughter contracted diphtheria. For fifteen days he kept vigil at her bedside. She recovered, but shortly after writing to inform his friends of the happy news, it turned out that he himself had been infected. He had no illusions that he would recover. Five days later, he died at the age of 49. In 1893, on the recommendation of a colleague, the syndrome he had described was given his name.

Two years after presenting the case of Mademoiselle X., Cotard decided to publish eleven more cases: eight women and three men, most of them middle-aged, whom he had observed in his own asylum over the course of many years.[10] They included Madame E., who believed that all her organs had been moved: kidneys, heart, lungs – nothing was in its usual place. A year later, she claimed that she no longer had a head or a body, that she was dead. If she moved, she would fall to pieces. She wouldn't accept any help, either: the slightest touch would make her splinter, like glass, into little bits. She died after a stay of fifteen years in the asylum.

Another woman was convinced that her family had been ruined, and that it was her fault: she expected to be arrested at any moment. She refused food on the grounds that she couldn't afford it, and she feared that her children would die of hunger. She believed she was suffering from a deadly disease and quarantined herself, because she didn't want to infect anybody. When she was finally persuaded to eat, she insisted on sitting at the servants' table because she felt deeply unworthy. She clung to these delusions right up to her death, eight years after admission.

Monsieur A.'s wife and son had both died within a short space of time, and he believed himself to be responsible for their deaths. He was the greatest criminal who had ever lived; he was the Antichrist. He wept and wailed that he wanted to die. At the

Maison de Santé in Vanves had its own park. This postcard from
the 1910s shows the island and its statue of Philippe Pinel,
who pioneered the humane treatment of the mentally ill.

same time he was convinced that his body had started to decompose. It contained no blood – cut me open, he would cry, then you'll see for yourselves. His heart had stopped beating, his head was an empty skull.

Yet another patient, admitted after a suicide attempt, continued to look for ways to kill himself. He demanded poison, a rope to hang himself, a knife to gouge out his eyes. He refused to eat and constantly confessed to imaginary crimes. He believed that his brain had turned to jelly and that he no longer had any testicles. '*Tuez-moi, tuez-moi!*' ['Kill me, kill me!'] he would yell for hours on end.[11] He wanted the staff to dig a hole somewhere and bury him like a dog.

It is perhaps the saddest group portrait imaginable. Most of the patients had been admitted after self-mutilation, hunger strike or suicide attempts. Besides believing they were dead, they suffered from deep depression, anxiety, suicidal thoughts and the feeling of being doomed. Some also thought they had a deadly disease. The delusion was chronic: most remained in the asylum for years or died there.

What all these patients had in common, Cotard wrote, was that they were *négateurs*, deniers: they denied literally everything. He himself called the syndrome '*délire des négations*', the 'mania of denial'. 'What is their name? They have no name. What is their age? They have no age. Where were they born? They were not born. Who were their father and mother? They have no father, no mother, no wife, no children.'[12] They no longer have anything; they no longer are anything. They are against everything; they want nothing. They have long ceased eating to still their hunger; they no longer feel hungry because they have no stomach and believe that everything they eat (with the greatest indifference) falls into a hole. Others eat only dry bread as a form of penance. They don't want to go to bed; they don't want to get out of bed. They want nothing more to do with their body: it no longer needs to be washed or cared for. It no longer feels like *their* body. One describes himself as 'a statue'; others refer to themselves in the third person. They resist help getting dressed or undressed, sometimes to the point of fighting tooth and nail. They themselves have lost all hope, but they also drive everyone around them to despair: there is nothing to be done with them.

The delusion of being immortal was something Cotard had come across before in patients with megalomania, but in such cases it sprang from delusions of grandeur. One of his patients believed his immortality to be a privilege conferred on him by Napoleon in 1804 (a quarter of a century before the man in question had even been born); another expected, when his time came, to be taken up to heaven in a fiery chariot like the prophet Elijah. Those who believed themselves to be dead, however, did not regard immortality as a privilege – rather as a curse: 'They exist in a condition that is neither life nor death, they are living dead.'[13] Had they been able to die, they would already be long dead. They felt doomed to live eternally, to atone for ever. Death could not relieve them of their suffering. Some patients came to the more or less logical conclusion that further suicide attempts were pointless and abandoned them – after all, you can't be deader than dead. Patients who suffered from religious mania, though, were inclined to try to kill themselves – despite believing that they were already dead or unable to die. They starved themselves, tried to set themselves on fire, stab themselves or end their lives in other appalling ways.

In words that reflected the other passion of his life – philosophy – Cotard wrote that, metaphysically speaking, people suffering from death delusions are nihilists; they deny everything: their organs, their bodies, their own existence.[14] Cotard didn't succeed in curing a single such patient. He ran up against the same brick wall as the daughter of the woman from Copenhagen, or the Amsterdam barber whose wife swallowed rat poison: the clearest proofs, the most convincing arguments had no effect whatsoever. Patients seemed to view reason as belonging to a world they were no longer part of. Cotard had nothing to offer them. Their situation was hopeless.

How I died

The Diagnostic and Statistical Manual (DSM) of Mental Disorders – the standard U.S. classification of mental disorders – does not list Cotard's syndrome as a distinct disorder. It categorizes the delusion of being dead or of lacking vital organs under the 'nihilistic delusions' that can accompany depression or schizophrenia. Furthermore, cases of disorders that aren't in the DSM mostly tend not to be recorded, so it's unclear how often death delusions occur. To date, around two hundred cases have been described in psychiatric and neurological literature. In 1995 two psychiatrists analysed one hundred cases.[15] In 2007 another 37 cases were collected from English-language sources, which largely repeated the conclusions reached in 1995.[16] The youngest of the one hundred patients in the former study was 16, the oldest 81. Most were in their fifties: Cotard's mainly strikes in middle age. It occurs as frequently in men as women. In nine out of ten cases, the patient is severely depressed. They nearly all suffer from nihilistic delusions about their bodies. Two out of three doubt their own existence, believe themselves to have a deadly disease or are racked by fear and shame. Only a minority – one in five – hear voices or have visual hallucinations. Six of Jules Cotard's eleven original patients believed they were immortal; almost exactly the same proportion was found in 1995: 55 per cent. Analgesia – the inability to feel pain that Cotard had noted as he repeatedly jabbed Mademoiselle X. with a needle – can lead patients to mutilate themselves horribly in an effort to feel *something*.

Patients with this syndrome want above all to be left in peace, but they rarely stop talking altogether. They continue to feel a need to explain why they don't eat, don't wash, don't want to see their family any more. A 66-year-old woman claimed that her stomach and oesophagus were glued together, and that she could therefore no longer digest food. Within the space of a month she lost 15 kilograms (33 lb) and would, like Mademoiselle X., have wound up starving herself had her daughter not had her sectioned.

A 59-year-old man likewise refused to eat, claiming that it was because his throat had been burned. His wife felt she had no choice but to have him committed to a mental hospital. Both patients responded well to antipsychotic medication and, within a month, were well enough to be released.[17]

Cotard's sufferers almost always have a 'story' about how they died, and it often ties in with a real-life event. One believed herself to have been poisoned with an emetic that had indeed been given to patients on her ward. Another woman, who had fainted on a flight from Britain to the United States, became convinced that she died at the time. A ten-year-old boy was sure that he had died in the Oklahoma City Bombing in 1995. A man who had been X-rayed believed that his organs had been irreparably damaged in the process. Interestingly, he was himself a doctor: he insisted on going over the X-rays with the medical staff treating him, pointing out where his organs were missing. Other patients believed that a failed suicide attempt had in fact succeeded, or that vital organs had been secretly removed during an actual operation.

Patients with this syndrome are completely immune to evidence to the contrary. During an eye examination, a man who had been admitted to hospital for depression claimed that he had no eyes.[18] Even when placed in front of a mirror, he stuck to this claim, saying that a doctor had removed his eyes when he was in A&E (the accident and emergency department). A little later, during a heart examination, he told a similar story to the cardiologist: his heart, too, had been removed in A&E. He politely declined the offer to listen to his own heartbeat with the stethoscope; it would be pointless.

Cotard's syndrome often goes hand in hand with depression, but the depression isn't a consequence of Cotard's – it is there before the delusion develops. Having a death delusion exacerbates existing feelings of anxiety and black thoughts. Many sufferers worry about what should be done with their bodies. They are dead, so their corpse will putrefy, will start to rot and stink – it probably already stinks – and needs to be buried or burnt as soon as possible. Somebody must do something. But nobody is helping.

The deadly disease that patients believe themselves to be suffering from varies. In Cotard's day it was often syphilis – not because patients really had it, or had reason to think they were infected, but because it tied in with their delusions of guilt, sin and punishment. They had brought death upon themselves through their wicked lifestyle. In the 1980s, its place was taken by AIDS.[19]

Sufferers' notions about where they are and what will become of them are shaped by the age and culture in which they live. In Cotard's day, his patients thought that they were in hell, or doomed to hover eternally somewhere between life and death. Today, many sufferers believe that they are zombies, or the 'walking dead'. Some psychiatrists have adopted this terminology and call Cotard's the 'Walking Corpse' syndrome.[20] If you Google images of Cotard's syndrome you're treated to a creepy barrage of putrefying zombies: half-decayed bodies doomed to eternal life. These undead are also to be found in psychiatric clinics. A man with Cotard's who was sectioned after committing an assault stated that he had drowned in a lake many years earlier and had then been reanimated with radiation from mobile phones to become a zombie.[21] He felt no remorse about attacking people because they were zombies too, and therefore already dead.

Organic disorders

In the 37 case studies of Cotard's that included details of neurological research, the delusion turned out to occur in combination with a slew of organic disorders: strokes, tumours, epilepsy, migraine, multiple sclerosis, dementia, Parkinson's and brain damage resulting from trauma. But the picture this presents is somewhat misleading. It is true that more brain abnormalities are detected in scans of people with Cotard's than in a control group of people without the syndrome. However, most Cotard's sufferers do *not* have brain abnormalities. On the whole, their EEGs are normal. In addition, the authors of this analysis themselves warned that the number of organic disorders had almost certainly been inflated by overreporting.[22] Cotard's is a relatively rare syndrome – but not so rare that articles about 'ordinary' cases will be accepted for publication. To be accepted, a paper must have one or more peculiarities to report. Articles about Cotard's are always about 'Cotard's *plus*': Cotard's after a haemorrhage on the right side of the brain, Cotard's after a typhus infection, Cotard's after a brain operation, Cotard's after a road accident.

Cotard's can also take the form of an acute variant. In 2007 a 35-year-old woman was brought into the casualty department of a Stockholm hospital.[23] She could hardly walk, was crying hysterically and was so upset that she couldn't say what was wrong, though she seemed to be terrified of something. Her mother said that she'd been like that for several hours. She had previously had a kidney transplant, but it had failed, and she now had to come to hospital for regular dialysis. After developing herpes, she'd been given an anti-viral medicine. A few days after starting to take it, she became restless at night, and a day later she said that her body no longer felt familiar, she felt cut off from the world. The next night she started to hallucinate and became fearful. Still crying, the poor woman was hooked up to the dialysis machine.

About three quarters of an hour later she calmed down somewhat and could explain what had happened. She had panicked, she explained, because she thought that she was dead. That feeling had now ebbed away, but she'd spent the previous hours in a state of extreme terror. An hour later, she said that her arm felt strange – when she touched it, it seemed like someone else's arm. The treatment with anti-viral medication was stopped. Within a few days, she felt perfectly normal. A 36-year-old man who had undergone a bone marrow transplant also developed a herpes infection and was given the same anti-viral medication. He had woken up in terror, believing that he was dead. The delusion vanished as soon as the medicine had passed out of his system. The researchers concluded that the drug, combined with renal failure, could lead to an acute form of Cotard's. Such patients should therefore not be referred to the psychiatric wing but given dialysis as soon as possible.

In 2018, researchers at the Mayo Clinic in Rochester, Minnesota, reported on eleven patients with Cotard's. They included a 74-year-old man who claimed that he had been stabbed to death in the care home where he lived and a 39-year-old woman who maintained that hospital staff had killed her with an overdose. Just like the elderly lady in Copenhagen, she expressed a wish to sleep 'in my coffin.'[24] A CT scan or an MRI scan had been done in the case of all eleven patients. The findings reflected what had been thought for some time: *if* there is any damage, it usually takes the form of abnormalities in the right side of the brain, primarily in the frontal lobe and the temporal lobe. The abnormalities varied in nature

– loss of volume, infarcts and damage to vessels – but whatever the issue, the damage was predominantly on the right side.

In Cotard's day, there was no effective way of treating death delusion. The prognosis was bleak: a permanent stay in a psychiatric institution under suicide watch. Cotard's is still a serious condition today, but with better prospects than one and a half centuries ago.[25] If the death delusion is part of a depression that responds well to treatment, it will usually disappear, sometimes quite abruptly. In the case of delusional disorders, antipsychotics will often dispel the delusion. In each case, treatment targets the underlying disorder, not the delusion itself. The fact that the delusion usually occurs in cases of very severe depression is reflected in dozens of articles in which electroshock therapy is presented as the most effective treatment. A single case of spontaneous recovery occurred after two epileptic fits. Incidentally, the literature on treatments is as anecdotal as that on attendant organic disorders, and it suffers from the evil that articles on successful treatment are more likely to be accepted for publication. Failed treatments – just as informative – do not make the journals. The fact that in 2010 a 59-year-old sufferer in Taiwan was cured of her delusion through a combination of antidepressants and antipsychotics does not tell us very much as long as the number of patients that did *not* respond to those drugs remains unknown.

Phantoms and doppelgängers

Psychiatric literature is bursting with bizarre delusions. People with De Clérambault's syndrome or erotomania (usually, but not exclusively, women) believe that a high-ranking or famous person is secretly in love with them. They make great efforts to contact their 'admirer', in ways that can be threatening. Others claim to be Napoleon or that they were stolen from a royal cradle as a baby. Delusions are persistent, and attempts to correct them are often inventively woven into a delusional system. If the person supposedly 'in love' with the De Clérambault's sufferer rejects their attempts at contact, for example, it's not because their love is not returned, but because the loved one is married and so not free to follow their heart.

Delusional systems are often cleverly constructed and consistent, even if their premise is bizarre. Some delusions could – hypothetically – be true. The famous guitarist on whom the woman with De Clérambault's is fixated *could* be in love with her; there *could* be listening devices hidden in wall sockets. But believing yourself to be dead is another kettle of fish entirely, raising all kinds of philosophical questions. How do Cotard's sufferers erase *themselves* from their experience? Who is the 'I' in 'I am dead'? What kind of neurological or psychological damage can prompt the inherently contradictory experience of thinking that you are dead? The beginning of an answer only dawned when the mystery was made greater. (In neurology and psychiatry, it is sometimes worth adding a puzzle or two.)

The first puzzle concerns phantom sensations: feelings that seem to come from a limb or other organ that is missing. Daniel Heller-Roazen, professor of comparative literature at Princeton University, wondered whether these sensations might be the mirror image of Cotard's syndrome.[26] After all, people who think their amputated leg must still be there because they can feel it are reasoning just as logically as Cotard's poor patients, who concluded from the absence of any feeling that they were dead. Phantom sensations arise because the brain area that represents the limbs is still active and intact but hasn't adapted to the loss.

The brain is working with an outdated map of the body. As a result, a sufferer feels as if they are still whole. In the case of Cotard's syndrome it may well be the other way round. The neurological representation of the body may have stopped functioning or be inaccessible. The resultant lack of sensations makes people think that parts of their body are missing or have died off.

In the case of more extreme neurological damage or injuries at other sites, the patient starts to think that a whole range of body parts – eyes, liver, heart, stomach – have disappeared, and that the loss of these vital organs means they must be dead.

In 1897 the French psychiatrist Jules Séglas recorded the case of a female patient who gave an extremely lucid description of the changes she had felt in her bodily perceptions when she thought she was dead:

I no longer have the same bodily sensations. I've felt my head change shape at least ten times, I no longer have a brain. My head and bones seem to be made of wood, I don't feel them like I used to. I no longer have a heart: there is something beating where my heart used to be, but it isn't my heart, it doesn't beat in the same way as before. I no longer have a stomach either, I never feel hungry anymore. When I eat something I can taste it, but when it's in my throat I don't feel it anymore, it's as if it just falls into a hole. Before, I used to feel it going down to my stomach, whether it was hot or cold. I can no longer feel my eyes moving, to move them I have to turn my head. Before, when I cried, I could feel my heart lurch and that made me feel better, now I cry without feeling anything, I've no idea how this has come about.[27]

Patients with phantom sensations live with the ghosts of limbs; Cotard's patients make themselves ghosts.

There is a second, equally puzzling condition that skews a person's perception of identity – but in this case someone else's. In 1923 the Parisian psychiatrist Joseph Capgras published a case study of a 53-year-old woman who claimed that her daughter had been kidnapped and replaced by a *sosie*, an imposter who looked identical.[28] This doppelgänger delusion only ever involves those closest to a patient: their partners, close relatives, dear friends. Sufferers from Capgras believe that their loved ones have been replaced with imposters who look exactly like them. They are actors, robots or – more recently – aliens. Capgras syndrome is usually linked to organic damage and can be sparked by a whole series of neurological diseases and disorders, from Parkinson's and multiple sclerosis to epilepsy and migraine. Roughly one in ten people with Alzheimer's have Capgras-like symptoms, which means that family members have to deal with the added distress of being taken for an imposter. There is no correcting this delusion. People try all sorts of ways to prove their identity, for example by telling the patient something that only they could know. But this inevitably backfires: the patient, already suspicious, begins to think that their vanished loved ones are in on the plot – how could the imposter otherwise have got access to this knowledge?

Sufferers sometimes become aggressive, taking out their worries and frustration on the doppelgänger, who in one case paid with their life. In Missouri an 82-year-old man was bludgeoned to death and decapitated by his stepson, a Capgras sufferer. The man was sure that he was dealing with a robot and scoured the inside of his stepfather's head for batteries and microfilms to prove it.[29]

Capgras sometimes occurs in combination with Cotard's. The associated neurological damage might explain perceptions like those experienced by Séglas' patient. Incidental cases had already been documented long before Cotard and Capgras arrived on the scene. Jacobus Schroeder van der Kolk, a doctor from Utrecht, wrote in a manual (published posthumously in 1863) that his patients included a young, caring mother.[30] After a traumatic event – her six-month-old baby had died of fits in her lap – she developed delusions. It was as if nothing really penetrated through to her any more. She felt guilty about that lack of feeling, and after a time felt sure that she was dead, expecting 'to be placed in a coffin any minute.' She could hardly be persuaded to eat: 'after all, she was dead, so needed no nourishment.' Then she developed yet another delusion. She became convinced that her husband and sisters had died, and that it was her fault. The fact that they still visited her must mean that she was being deceived: 'the persons who purported to be them were in fact evil spirits, who had dressed up in the clothes of her husband and sisters in order to torment her.'[31] Cotard's patients included a woman who no longer recognized her husband and children; she believed that the woman who claimed to be her daughter was really a devil in disguise.

Psychiatric literature features occasional reports of patients suffering simultaneously or alternately from Capgras and Cotard's. The elderly man in the group of patients from the Mayo Clinic who thought that he had been stabbed to death also believed that his wife had been replaced by an alien. Another well-documented case is that of the 35-year-old H., a man with severe clinical depression who was sure that his family was plotting to kill him.[32] When admitted, he was fearful, suspicious and agitated. Neurological examination revealed no abnormalities. After a while he was discharged, but he was soon brought back in an ambulance, having stabbed himself in the arm with a kitchen knife to prove that there was no blood left in his body. He told everyone that he was dead. Treatment with chlorpromazine, an antipsychotic, dispelled the delusion. When the man had calmed down a bit, he explained that he had thought he was dead because of 'feeling nothing inside'. This belief was strengthened by the fact that he no longer recognized his familiar surroundings and family. An EEG scan revealed epileptic activity in the temporal lobe, and he scored poorly on a test for recalling faces. Otherwise, his memory was normal. He was again discharged. Six months later he was back again, this time with paranoid delusions: he believed he was being followed by white cars and that he was being talked about on the radio and television. The death delusion had gone. But now his father had been replaced by a doppelgänger.

A similar pattern of neurological damage and psychiatric symptoms was found in the case of a 28-year-old stockbroker from Scotland who suffered serious damage to his right temporal lobe and the surface of both frontal lobes in a motorbike accident.[33] He, too, complained of a sense of alienation: walking through a street in Edinburgh that he knew well, he thought that the original houses had been knocked down and replaced by buildings that looked similar but were very clearly different. He also performed poorly on a memory test for facial recognition. After he recovered, his mother took him on a trip to South Africa. There he developed the delusion that he had died many years before, probably of an infection or AIDS, and that he was now in hell – which was why it was so terribly hot. He was being accompanied in Hades by the spirit of his mother, who he believed to be still in Scotland, asleep.

This alternation or combination of Cotard's and Capgras in one and the same patient suggests that both delusions might be triggered by the same pathological process. At present, the most convincing explanation for the doppelgänger delusion is that the brain area responsible for visually recognizing a loved one is still intact, but that the information no longer reaches the brain area that generates the associated emotional response. The patient sees his wife come in, but to his astonishment and shock he doesn't feel the same emotional warmth as before. So the patient is forced to conclude that this woman *cannot* be his wife, even though she looks exactly like her. This mental tussle had already been noted back in the day by Joseph Capgras, who spoke of a conflict in the *logique des émotions* (emotional logic). This is why patients suspect only people whom they love of being doppelgängers: in encounters with others, the absence of an emotional reaction is normal.

Patients with Cotard's experience an even more drastic disturbance to their emotions: they no longer feel anything at all. Whatever happens to them, whoever visits them, it all leaves them completely cold. But whereas patients with Capgras explain this internal numbness by making adjustments in the outside world – hardly surprising I feel nothing, since she isn't my real wife – Cotard's sufferers look for the cause in themselves. They feel nothing because the organs with which they could feel – their heart, their blood, their intestines – have disappeared or, worse still, because they are already dead.

The hypothesis about the neurological explanation for Capgras was formulated in 1990 by two cognitive neuropsychologists.[34] Inspired by the notion that something similar might apply in the case of Cotard's, another study uncovered yet more parallels.

For example, people with Capgras and Cotard's scored equally badly on facial and facial expression recognition tests, whereas all other memory functions proved fully intact.[35] The list of accompanying disorders also overlapped. In 2013 the first PET scan was performed on a patient with Cotard's, in the hope of pinpointing exactly which areas of the brain were affected.[36] The case was a particularly poignant one.

Fried brains

After two divorces, Graham Harrison, a 48-year-old installer of water meters from Exeter, became severely depressed. One day, he tried to end his life by pulling a hairdryer into the bath. The attempt failed, and he came round in hospital. From then on, he believed that he had 'fried' his brain and that he was brain dead. He could no longer taste or smell anything, lost interest in the car he formerly idolized and stopped looking after himself. If his brothers had not made sure that he ate, he would have starved to death. In 2013 science journalist Helen Thomson came to interview him for a *New Scientist* article entitled 'First Interview with a Dead Man.'[37] He told her, 'I didn't really have any thoughts – no emotions. I didn't feel anything ... I didn't even remember what pleasure was like. I just had this blank mind and I knew – I couldn't say why – I just knew that I didn't have a brain no more.'[38] Graham admitted that he could still see, think and speak, and realized that that proved he was alive – but how that was possible with a dead brain was something he couldn't explain. He kept quiet about how he felt. It was pointless to tell people that he was dead; they would think he was crazy. 'No, I didn't tell anybody. It's kind of a weird thing to say to someone – "I ain't got no brain."'[39] He considered killing himself by putting his head on a railway line but had decided against it. 'I'm sure my head would still be there, I'd still be able to speak because I'm already dead, so the train can't really kill me.'[40] Medication was tried, but it didn't help. He told his GP that he wanted to prove that he was brain dead, and a neurological clinic in Liège agreed to examine him.

Graham underwent various neurophysiological tests. Despite being 'brain dead', his EEG showed no abnormalities. However, the PET scan revealed something surprising. Metabolic activity across large areas of the brain was so low that it resembled that of someone who'd been anaesthetized or was deeply asleep. The frontal and parietal lobes were the worst affected – precisely the areas thought to be vital to core consciousness and our ability to integrate experiences into a 'self'. It is these areas that are most deactivated during loss of consciousness or an epileptic fit.

It's a pattern that points to delusion being caused by a combination of two factors. First, damage to the temporal lobe prevents patients from receiving sensations either from external stimuli or from within their own bodies. No internal signals are coming in. It's as if every single organ or body part – heart, stomach, gut – has blacked out. That incomprehensible inner silence makes patients conclude that either their organs are missing or dead, or that they themselves have died. Initially, that's just a bizarre thought that pops up but is soon rejected. It only becomes a real delusion if there is damage to the frontal lobes. Because that's the second factor: the prefrontal cortex is no longer able to correct bizarre ideas. Without that correction, an absurd notion can become a fixed belief. According to this theory, a lot of people could potentially believe that they were dead – but the idea remains a fleeting thought, because their frontal lobes are still functioning properly. It's only when the censor at the front of the brain is silenced that the way is paved for delusion, and a person will sooner or later be diagnosed as suffering from Cotard's. (Cases where the patient has no damage to the brain will be discussed below.)

I think, therefore I am

Sartre's story 'The Wall' is set in the Spanish Civil War. It's nighttime; three men are sitting in a cell. They are to be executed at dawn. One of them tries to imagine what it will be like, being dead. He can't:

> I'm a materialist, I swear it; I'm not going off my head. But there's something not quite kosher about it. I can see my corpse; no problem there, but it's *me* seeing it, with my eyes. I'd need to get my head round the idea . . . the idea that I won't see anything again, or hear anything, and that the world will continue for everyone else. We're not designed for ideas like that, Pablo.[41]

The fact that you can't mentally remove yourself from an observation, an idea or an act of reasoning is a Cartesian axiom. The ability to doubt everything except the act of doubt itself – because doubting proves that you exist as a rational being – was for Descartes the premise of his 'I think, therefore I am.' It was the clear, self-evident foundation from which the rest of his philosophy sprang – a logic that no reasonable soul could doubt. Even in the darkest depths of depression, most people remain sufficiently rational to believe in their own existence. In his blackest hours, Rimbaud could still invoke Descartes. In 1873, in his prose poem *A Season in Hell*, he wrote: 'I believe I am in hell, therefore I am.'

But the land of Descartes, Rimbaud and Sartre also brought forth Jules Cotard. His patients lived with their own axiom: 'I am dead, even though I think.' From that cornerstone, they constructed their own philosophy: how this had come about, where they were now, what their fate would be. To them, their deadness was as incontestable as his existence as a rational being was to Descartes.

If Descartes' axiom expresses reason, rationality and proof, the axiom of a Cotard's sufferer could be seen as embodying the opposite: illusion, delusion and self-deception. But to take that black-and-white approach would be to oversimplify. It allows no scope for forms of reasoning that seek to preserve consistency within a delusion – both between its different elements and between reality and what a sufferer experiences and believes. Perhaps the best analogy for this type of reasoning was the one given in 1867 by the German physiologist, mathematician and philosopher Hermann von Helmholtz in his monumental *Treatise on Physiological Optics*.

When you look at someone walking away from you, within a few seconds, the image of them projected on your retina is reduced to a fraction of what it was. Yet you perceive their size as relatively constant: you don't see them shrink as they walk. This 'size constancy' is the result of a mechanism that's automatically triggered as soon as your visual apparatus signals that something or someone is moving away from you. The shrinking size of the image on your retina is multiplied by indicators of distance, and this happens so quickly and accurately that the resulting image remains more or less constant. These calculations are the product of rapid interactions between your eye and the primary visual cortex at the rear of your brain. It's a mechanism you can't switch on or off: you don't have access to all that multiplication (or division, when an object or person is getting nearer). Size constancy, Helmholtz wrote, is the result of an '*unbewusster Schluss*', an unconscious inference.

Once this mechanism is fine-tuned – and this happens back in early infancy – it just gets on with its work without us being aware of it. But it can be manipulated, either in natural circumstances or experiments. When you see a full moon rising low on the horizon at the end of the street, because of the many indicators of distance – pavements, houses, trees – you perceive it as very much bigger than the little moon that sails through the clouds directly above your head a few hours later. Both moons are projected on your retinas as exactly the same size, but then something goes wrong when multiplying these retinal images. In the case of the moon in the sky, the multiplying factors that make the moon at the end of the street so huge are lacking. What you see is still the result of an unconscious calculation, but this time you're being misled by your own sensory system. Your brain has been calculating with faulty data. Dozens of optical illusions are based on manipulating this mechanism, which in normal circumstances maintains size constancy with perfect accuracy.

Being aware of what causes an illusion doesn't make you any less susceptible to it. Even after the moon illusion has been explained, the moon at the end of the street remains what it was: surreally big. The rational knowledge that you're overestimating its size is divorced from your perception. Is this what it's like to have a delusion? Your brain has done some reasoning and come up with a conclusion that seems obvious and incontestable. So there's nothing for it but to try to make it fit with your actions and experiences.

In the case of Cotard's, some similar unconscious mechanism is at work. But instead of visual images, it bases its calculations on a different kind of input: the internal deadness, the absence of sensations and the numbed emotions. The outcome of the sum is: I am dead. No doubt about it.

What types of organic damage can cause this illusion? The optic nerves, routes and brain areas that collectively ensure size constancy are precisely known, and damage to specific parts of that trajectory have predictable consequences. But that isn't the case with Cotard's. For a start, some people with this condition don't have any detectable organic damage. (To argue that there might be damage, but that it just can't be detected with the current instruments, is clutching at straws.) Moreover, where damage is established, it proves to be highly diffuse. Vascular damage, strokes, atrophy, tumours – Cotard's is associated with the most diverse kinds of injury. Nor does the location of brain damage make the picture much clearer. Often, functions in the temporal and frontal lobes or connections between the two are disrupted, but since this is such a large brain area, the actual mechanism involved is still a mystery. For now, theories and research focus on the fact that Cotard's more often features disorders in the right side of the brain. The same applies to Capgras, as well as to feeling that a hand or leg doesn't belong to your body. The fact that some patients suffer simultaneously or alternately from Cotard's and Capgras could point to an overlapping disease process.

Helmholtz's idea of an unconscious process of deduction conjures up associations with cold logic and calculation. But that's as far as the analogy goes. Cotard's syndrome is linked to a terrifying sense of crisis and alienation. Each patient expresses this in their own way. The feeling of no longer being one's former self, housed in one's familiar body, prompts some to speak of themselves in the third person. Others felt, as Jules Cotard explained, like a statue or a machine (these days an automaton or a robot). The body feels different; it's no longer made of flesh and blood but of glass, stone, wood, steel or iron. It no longer lives, it is hollow, the veins are empty, the throat is just a hole, the organs have disappeared, it's a carcass. But however absurd the conviction, the *perception* is very real; the patient feels what they feel, even if that is the absence of feeling. Unconscious inferences are binding.

Not long after Helmholtz's death in 1894, Sigmund Freud introduced an entirely different type of unconsciousness. This, too, had its own laws and mechanisms. Shortly after the outbreak of the First World War, Freud published his reflections on war and death. Rationally speaking, he wrote, everyone acknowledges that life ends with death: 'Everyone of us owes nature his death and must be prepared to pay his debt.'[42] But that is not how we live and act, and that's because one's own death is unimaginable. If you try to imagine your death, you remain present as a spectator. Therefore, Freud continued, 'no one believes in his own death, which amounts to saying: in the unconscious everyone is convinced of his immortality.'[43] Isn't *that* precisely what's so terrifying about the Cotard's delusion – that even this conviction can be taken away from you?

THREE

The Murder of the Widow Van Sandbrink

The Dutch town of Amersfoort was formerly the centre of a flourishing tobacco-farming region. Tobacco fields, bordered with hedges to protect the vulnerable plants from wind, carpeted the southern slopes of the hills around the town. The cultivation of tobacco had started in the mid-seventeenth century and, by 1680, there were around two hundred tobacco farmers in the vicinity of Amersfoort. The crop was hung to dry in ventilated barns or in church attics. Much of it was ground to produce snuff tobacco. The nearby cities of Amsterdam and Utrecht were big sales areas.

Around 1800 the tide began to turn. There was less demand for snuff tobacco and increasing competition from Kentucky and Virginia in the USA. In the decades that followed, one after another tobacco farmer was forced out of business. They included Lucas Kier van Ootmarsum, known locally as Kier.[1] In 1814 he had become a partner in his father's business, a merchant in spirits and tobacco pipes. He got into tobacco farming later, at a time when the sector was already struggling. The venture did not end well.

In 1810 Kier had married Cornelia van Leeuwen. The couple went on to have twelve children, and their shrinking income wasn't enough to feed ever more hungry little mouths. Luckily Cornelia had an aunt who was concerned about the family's welfare. Christine van Sandbrink was a widow who had no children of her own. Ownership of a taproom had made her extremely wealthy: she possessed eight houses in Amersfoort, as well as a substantial amount of cash and a government bond. On top of that, she was owed 9,000 guilders – back then, a vast sum of money.

In May 1826 Kier had borrowed 4,000 guilders from 'Auntie', putting up the harvests of the past three years (currently drying nearby) as collateral. But the cash injection did not help. In September of that same year, Kier was forced to close the business. A merchant snapped up his entire stock, including two horses, a cart, a waggon, a plough, bales of hay, sacks of oats and

Tobacco farms on both sides of Hogeweg, on what was then the outer border of Amersfoort. The Lazarushuis (on the right), built around 1410 for the care of lepers, had been converted into a tobacco farm. The tobacco barn on the left is of a type characteristic to Amersfoort, with an asymmetric roof. Engraving by Paulus van Liender, 1759.

rye, planks, wheelbarrows, baskets, roof tiles, racks as well as whatever miscellaneous items were to be found on the farm.

Initially, Christine van Sandbrink had named her niece and husband as her heirs in her will. For a while, Kier had managed to keep his creditors at bay with stories about the vast amount of money he would be inheriting. But Auntie was no mug. She was well aware that her loans to Kier were simply being swallowed up by his ballooning debt, which meant that, when she died, the money would go to his creditors. In 1833 – by then she was 68 – she decided to change her will. Instead of Cornelia and Kier, their children would be the beneficiaries. When Kier heard this, he apparently lamented 'Now I'm done for!' After that, his relationship with Auntie quickly went downhill.

Early in the evening of 17 August 1835, Kier rang the doorbell of her house in Amersfoort's upmarket Langestraat. The widow, who'd just been drinking coffee with a visitor, Jannetje, the sister of her maid Gijsje, let him in. Kier had already been round a few weeks earlier, and an argument had erupted that had nearly got out of hand. Auntie had reproached him with neglecting his family and of being workshy. Kier had got so angry that he'd struck the table. Shocked, she had got rid of him as soon as possible. As she saw Jannetje out, the young woman offered to stay with her for a while, but the widow didn't think that necessary. She wasn't worried; Kier seemed calm enough. He had sent Gijsje on an errand for him, and he was now alone with Auntie. At some point she walked to another room at the back of the house. As she did so, Kier followed her, drew a pistol and shot her in the head at close range. He then walked back home, sat next to his wife on the bench in front of the house and calmly awaited events.

The widow was found by Gijsje. In the gloom of the corridor, still filled with gun smoke, her hand had brushed against the face of her mistress. Her screams of 'Murder! Help! Oh God!' brought neighbours running. One of them, a basket weaver by the name of Meester, hurried to Kier's house in the nearby Sint Andriesstraat, to bring the bad tidings. Kier walked back to Langestraat with him, asking a bystander: 'What on earth has happened?' But a woman who lived close by had seen him coming out of the back of the house, half stumbling in his haste, just after she heard the shot. Kier was arrested. He confessed almost immediately, during his first interrogation.

The Kier case was documented in detail at the time, including witness statements, and in 2008 the historian Ignaz Matthey published an in-depth reconstruction of the murder and trial. It emerged that Kier had been planning to kill himself for some time, and he had set out his reasons in lengthy suicide notes. Letters and witness statements painted a picture of severe mental disturbance. Kier believed that his body was entirely hollow, that he had vomited up his heart, lungs and liver. He claimed that he had been dead since August 1834, and that his body was now made of iron and steel. In short, the dossier suggested a delusional complex corresponding with what is now called Cotard's syndrome.

Lucas Kier's case isn't the earliest of Cotard's pre-Cotard, but it is the first to be well-documented. Equally noteworthy is the fact that, to date, no other suicidal person with a death delusion has been recorded as committing a capital crime. But there is a third reason for taking a closer look at this historical case. Because what was the fate of someone who showed symptoms of Cotard's nearly half a century before the syndrome entered psychiatric literature and long before it impacted the practice of forensic psychiatry?

Was Kier – in 1835 – ill or guilty? Mad or sane? Should he be treated or punished? What was the appropriate destination for him: the gallows, prison or a madhouse?

The Second Death

After his arrest, an investigation took place. It was, in essence, a reconstruction of Kier's medical history. Through interviews with people who knew him – relatives, former colleagues, neighbours, the pastor, his doctor – the authorities tried to build up a picture of his mental state before and during the crime. It gradually became clear that Kier had been deranged for some time, at least a few years prior to the murder.

After his tobacco farm failed, he had been employed by the municipality of Amersfoort as an income tax assessor. Over the course of 1834, his colleagues saw signs of 'insanity' and said that he occasionally 'lost his wits'. He was fired. This was the start of a period of apathy – hence Mevrouw van Sandbrink's accusations of idleness.

Three years earlier, Kier and his wife had been struck by a terrible blow: within the space of three days, their four-year-old son and five-month-old baby died, probably of cholera. The measures taken by the authorities to prevent the disease from spreading made their loss even more traumatic. After the children's deaths had been certified, local officials placed the little bodies in coffins, sealing the seams with pitch. The house was fumigated with hydrochloric acid, and the children's bedding was burned.

Their family physician, Dr Van der Leeuw, reported that Kier had brewed some kind of anti-cholera potion and given it to others. Kier believed he was acting on higher instructions: he said that the medicine had been revealed to him 'in a dream by our

Lord'. He had been to see the doctor several times in 1834. Each time he 'gave long rambling speeches, always coming back to the same point, complaining of a persistent anxiety, sometimes claiming that he was entirely hollow, that everything inside him had decomposed, sometimes that he had spewed out his heart, lungs and liver.'

The investigators also called on his neighbours. Mevrouw De Jager said she avoided him as much as possible, but that on one occasion he told her that 'he had lost his heart, lungs and liver and had nothing left but muscles'. After that he had stood very close to her, saying that 'he wanted to take a good look at her – it being such a strange world, and all the people in it being dead.'

Another neighbour, a weaver by the name of Moesman, stated that, on the whole, Kier talked sense, 'but often mixed with crazy pronouncements, for instance that the world no longer existed, that he was in his second life, that humanity had died out in 1834 – though we still appeared to exist – that no more children or animals would be born, that trees would no longer bear fruit, that the stars were no longer in their usual place and that he was now made entirely of iron and steel, so that it would no longer be easy for someone to cut his throat.'

A week before the murder, Kier had dropped in on Moesman. He showed him a pistol, saying 'Look, I can rest easy now – I can die whenever I want.' Moesman, who'd fought at Waterloo in 1815, recognized the model and warned Kier to be careful: the gun was old and, if he fired it, it might blow up in his hand. But Moesman couldn't persuade him to get rid of it. Kier said he planned to load it with two bullets – a terrible idea, Moesman thought – but Kier wasn't to be dissuaded. As he put it, 'That's better for splitting the head in two.' The day before the murder, Kier went to see a former farmhand of his, raising the subject of the end of the world again and gesturing that he intended to shoot himself.

This tense state of affairs now apparently received a fatal little push – an 'exciting cause', as Pinel would say, or trigger, as it would today be called. There was a problem at home. Kier was at loggerheads with Geertruida, his oldest daughter. She wanted to marry Fokke Kingma, a teacher from Friesland. At first, Kier refused to consent to the match. Later he relented, but he and his wife did not go to her wedding. The day on which Geertruida exchanged vows with Fokke in Leeuwarden was the day on which Kier killed Christine van Sandbrink. Later he confessed that he had left the house in order to shoot himself in the head, but according to his own peculiar reasoning, he had first wanted to spare Auntie 'death by martyrdom'.

After Kier's arrest, a writing desk was brought from his house to the prison and the drawer opened in his presence. It contained three suicide notes. As a rule, such missives are hastily scrawled and brief.[2] But a month before his intended suicide, Kier had written down in detail to his family what he had been unable to say to them. In a long letter to his wife and children, he referred fourteen times to texts from the Book of Revelation, the Bible's book of prophesies about the Last Judgement.[3]

Around 1830, many Christians believed that the Second Coming was imminent. Some prophesied that the world was about to end. In the Netherlands, the best known of these prophets was Hendrik Hentzepeter, doorkeeper of the Mauritshuis picture gallery in The Hague and author of eighteen pamphlets about the apocalypse. In 1832 he predicted that the Second Coming was at hand: 'The evening of the great Saturday has drawn near, the week of the world is almost at an end.'[4] Kier also thought the end was nigh: he believed that the Last Judgement had dawned in the spring of 1834. Bolts of lightning and violent earthquakes had hurled the countries that were 'under Papal and Turkish sway' into 'the bottom of the abyss', he wrote, 'and we too along with them'. The catastrophe had taken place as the Earth revolved, so the stars were now on the wrong side of the firmament – as he had recently observed with his own eyes during a visit to Amsterdam. Everyone was now in a state of Second Death which in the Book of Revelation stood for eternal damnation. He believed that his own body had died the previous August. Working to support his family was no longer possible. The thought of eating repulsed him. He couldn't sleep either, being tormented by hellish dreams. He no longer wished to live in the fake world that had replaced the real one after its destruction.

Kier had also left letters for his father and brother Jacobus. After a farewell to his brother – 'Goodnight dear ones! For ever, goodnight!' – he asked him to do one last thing for him: 'Please make sure that the nerve in my neck is cut right through, so that I am as dead as can be, and won't be eternally tormented by any thoughts that might linger in me.'

Was he faking madness?

Kier's trial took place two months after the death of the widow Van Sandbrink. Nineteen witnesses were heard. During the preliminary investigation, one of Kier's interrogations had to be stopped because he answered every question with 'I can't remember anything.' Suspicions arose that he was malingering. The prosecutor, a Mr Provó Kluit, claimed that in the months before the murder Kier had been deliberately saying all the bizarre things reported by witnesses so as to escape punishment on the grounds of insanity.

During the trial, the prosecutor maintained that, despite all the witness statements, the suspect had on the fatal day apparently been lucid enough to plan and carry out this monstrous deed. He also alleged that Kier had feigned the 'confused state' he was supposedly in during his interrogations. This saddled the judges with a dilemma. If they accepted the defence of insanity, the case would have to be dismissed under Article 64 of the Penal Code. He could not then be forcibly committed to an asylum.[5] Apparently they regarded this as a bridge too far: on 31 October 1835, Kier was found guilty of 'wilful murder with intent'. He was condemned to the gallows.

In the months that followed, the authorities were deluged by appeals for a reprieve: at least sixty were received. The petitioners ranged from citizens of Amersfoort to members of the town council. In his appeal, the prison chaplain Van der Leeuw – tasked with preparing Kier for the hereafter – described him as 'a person who is unfeeling to an extent that is quite incredible', undoubtedly the result of 'lunacy'. In a joint petition, Kier's father, brother and wife pointed out that it had been clear for some time to a great many local citizens that he was no longer in full possession of his wits. His current state was proof of that: it was heartrending to see how a man formerly so full of feeling now 'viewed his fate and the misfortune of his relations with cold indifference'.

The appeal for clemency was submitted to the prosecutor for advice. Provó Kluit paid a visit to the condemned man in his cell but did not change his mind; he remained convinced that Kier was feigning insanity: this was 'an angry and dangerous man'. Kier seemed to him 'somewhat vacant, but extremely dangerous, insensitive and unfathomable'. He felt no pity at all for Kier, only for his family. So when he subsequently recommended that clemency be granted, his decision was prompted by 'their interest, not the criminal's'. Had the Van Ootmarsums not been such an upright family, he would have had no scruples about hanging Kier. The High Court committee that advised the monarch on appeals for clemency recommended that the death sentence be commuted to life imprisonment.

On the first page of their advisory report to King Willem I is a note pencilled by the king himself. Wouldn't it be better to put Kier in a madhouse? In the end it was prison. His last journey was to Leeuwarden, where he was incarcerated in the Blokhuispoort jail.

Retrospective diagnostics

The Kier case records don't mention an inability to feel pain or self-mutilation. But the delusion of being damned – as reported by Kier's doctor, neighbours and the prison chaplain, and confirmed in his suicide notes – ties in with symptoms that Jules Cotard noted in his patients with *délire des négations* in 1880. Cotard also often found a delusion of immortality coexisting with religious mania. As far back as August 1834, Kier must have felt torn between the belief that his body was a dead, iron shell and the fear that when he *did* die, he would be eternally tormented by his last thoughts if no one severed the nerves of his neck. The same long list of everything that no longer existed – from plants, animals, people and continents to the whole world and stars – is something that Cotard also heard from his *négateurs* and which he classified as nihilistic delusions. The conviction of lacking vital organs like lungs and liver, and the anxiety, depression and changing bodily sensations, map closely with the characteristics listed by Cotard as part of his syndrome. They also match the clinical picture that was distilled in 1995 from an analysis of one hundred case studies.[6] With the knowledge that we have now, the suspicion that Kier was feigning mental illness seems completely unfounded.

What was atypical was that he committed murder. People with Cotard's tend to be withdrawn and apathetic, though they can respond aggressively to attempts to force them to dress or look after themselves. Kier's motive for the murder is hard to rhyme with his delusions. In his confession he said that he had murdered the widow to spare her death by martyrdom, but he didn't explain why he'd singled her out. If this was an act of mercy, why not extend it to his loved ones? Did his conflict with her play a role? Why did he go home after the murder as if nothing had happened? Previously, he'd claimed he had gone out in order to kill himself. He had reserved the second bullet in the pistol for himself, hadn't he? Why did he just fire the one shot? Much remains unclear about his motives.

The fact that Kier showed symptoms that were later categorized as Cotard's syndrome is hard to refute. But to call him a 'Cotard's sufferer' would amount to retrospective diagnostics: the diagnosing of people who lived in a previous age with diseases or disorders that were not known in their time. Georges Gilles de la Tourette described the syndrome that would be named after him in 1885, Hans Asperger his syndrome in 1944; the claims that the famous lexicographer Samuel Johnson, who died in 1784, suffered from Tourette's, or that Newton, Van Gogh and Wittgenstein had Asperger's, are examples of retrospective diagnostics. The last decade has seen a flood of posthumous diagnoses, for the autistic spectrum in particular.

It is a popular pastime, typically engaged in by doctors and psychiatrists but frowned upon by medical historians, who warn that it tends to focus on similarities at the expense of variation and difference. To be convincing, retrospective diagnostics requires both sensitivity to the historical context and knowing how much weight to attach to contemporary sources. In the case of Kier, for instance, you could easily lose sight of the fact that – with the exception of the suicide notes – all the information we have now was collected by judicial officials for the purpose of criminal investigation. What's more, the statements were drawn up after the crime and were from witnesses who knew what Kier had done. Both these factors have added their own measure of selectiveness. The problem is that today's readers don't know what they don't know: many questions will not have been asked at the time, and many that were asked will have been left out of the picture.

Philosopher of science Ian Hacking has also expressed reservations about retrospective diagnostics.[7] When people are labelled as having a specific condition, like Asperger's, they react to that label. The resulting interaction between the label and the labelled can cause a shift in the characteristics associated with that label. This 'looping effect' takes place in a shared cultural and historical space that didn't exist in Samuel Johnson or Vincent van Gogh's day. That's why the claim that Newton had Asperger's is so misleading: at the time there was no context of theories, treatments and institutions that would allow for a meaningful discourse about Asperger's.

Scope for looping effects is greater in some psychiatric classifications than others. In the case of Asperger's, the Aspies for Freedom group campaigns to prove that their 'disorder' is not a pathological abnormality but an instance of neurodiversity. In the case of Cotard's, there's less scope for that: people with a death delusion do not organize themselves in patient groups, nor do they claim to be merely neurodiverse. But even so, Hacking's objection remains valid. When assessing the legitimacy of a retrospective diagnosis, the specific medical context – or more often the lack of one – is an essential element. The same applies to Kier's case. To call him a Cotard's patient would be to disregard the fact that, in the view of the judicial authorities, a patient was what he was *not*. The prosecutor claimed he was feigning mental illness. The judges believed he was in his right mind when he committed the murder. And although the doctor, the chaplain, Kier's family and neighbours did believe he was deranged, they too ultimately judged him in *moral* terms. According to Auntie Van Sandbrink he was 'idle', according to the chaplain his religious views were 'dark, rigid and absurd' and according to his family he viewed his own fate with 'cold indifference'.

This is not how one speaks of a patient. But Kier's contemporaries simply didn't have that option. The medical discourse on the pattern of action and thought he displayed was still a thing of the future. As were the experts, psychiatric hospitals and legislation that went with that discourse. Everything that half a century later, after Cotard's findings, might have been viewed through a forensic-psychiatric lens – the psychosis, the suicidal thoughts, the religious mania, the sense of being doomed, the death delusion, the apathy, the absence of feeling, the fear of continuing to exist after death – probably resulting in commitment to a psychiatric hospital, was in 1835 still dealt with purely as a criminal process. The trajectory followed by Kier could only have ended in a prison cell or on the scaffold.

So jail it was, in Leeuwarden – the worst prison in the country. Blokhuispoort was notorious for miserable living conditions and a high death rate.[8] For many prisoners, 'life' amounted to a death sentence in slow motion. Lucas Kier van Ootmarsum stuck it out for five years, dying in August 1841 at the age of fifty.

He can't have had many visitors. Two prison guards who came to the civil registry to report his death said that they had no idea whether he had any family.

FOUR

Phantoms and Illusions

George Dedlow gradually regained consciousness and looked around him. He was lying on his back in a log cabin. A group of soldiers, sitting round a fire, seemed to be drawing lots for his possessions. No one was taking any notice of him. As unobtrusively as possible, he tried to establish how badly he was hurt. Finding that he could use his left hand, he ran his fingers along his right arm. A bullet had passed right through his biceps, and he no longer had any feeling in his right forearm and hand.

As the pain started to surge back, he reflected on the turn his fate had taken. Less than a month ago, he had been studying medicine at Jefferson Medical College in Philadelphia. His father was a doctor with a large practice, and the idea was for them to become partners. But in April 1861 the American Civil War started, and he had no choice but to break off his studies. He signed on with an Indiana infantry regiment as assistant surgeon.

From one day to the next, life became hard and dangerous. Food was scarce, and the soldiers had to be perpetually on guard against enemy snipers. Worse though, were the diseases that broke out. Dysentery, yellow fever and malaria were more lethal than cannonballs and bullets. Most of the war dead did not fall in action; they succumbed.

It had been a shortage of medicine that had landed Dedlow in this plight: badly injured, in enemy hands. His commanding officer had ordered him to ride under cover of darkness to another unit, some 32 kilometres (20 mi.) off, to ask for quinine. Not far from his destination, he had run into an ambush. He heard a shot and felt a blow on both arms. That blow was the last thing he remembered. He had no idea how long he had been unconscious. Not long afterwards, he was taken away in a cart.

He was examined in a rebel hospital near Atlanta. There was an open wound in his left arm, but his right arm was in a worse state. The hand was red and felt as if it were being perpetually rasped with hot files. The pain was unbearable. He begged for morphine, but the doctor replied that they had none: 'You know you [the Union Army] don't allow it to pass the lines.'

When the doctor returned an hour later, he was accompanied by two aides. Dedlow was told that only an amputation could end the pain. As he turned on his left side, he asked which of them would give him ether. 'We have none,' was the answer. After a searing flash of agony felt in every nerve fibre, relief suddenly came: for the first time in six weeks, he no longer felt any pain and fell asleep even before the flaps were sown over the stump.

In August 1863 a prisoner exchange took place. After a brief period of leave, the one-armed Dedlow returned to his regiment. But less than a month later, in the Battle of Chickamauga, he was again seriously wounded. When he awoke, he found himself under a tree; a little way off, he saw doctors busy at an operating table improvised from two barrels and a plank. Two doctors finally came over to him, briefly inspected his wounds, exchanged glances and walked away. Dedlow asked an orderly where he had been hit. 'Both thighs,' he responded, 'the doctors won't do nothing.' Shortly afterwards, someone squatted down next to him and pressed a cloth over his mouth. He smelt the familiar odour of chloroform, breathing it in greedily.

When Dedlow recovered consciousness, he felt a sharp cramp in his left leg. He tried to rub his calf with his remaining arm but was too weak to manage it. He gestured to an attendant.

'Just rub my left calf,' said I, 'if you please.'
'Calf?' said he. 'You ain't none. It's took off.'
'I know better,' said I. 'I have pain in both legs.'
'Wall, I never!' said he. 'You ain't got nary leg.'
As I did not believe him, he threw off the covers, and, to my horror, showed me that I had suffered amputation of both thighs, very high up.
'That will do,' said I, faintly.

He survived this operation, too. A month later he was well enough to be transferred from the crowded hospital in Chattanooga to a hospital in Nashville. He'd only been there a few weeks when fate struck him a final blow. An epidemic of hospital gangrene broke out. The men were moved into the open air as soon as possible, but for Dedlow it was too late. The wound in his left arm was already infected. Only amputation could save his life.

It was the beginning of his life as a torso. He still possessed the stump of his left arm, but it proved too tender to bear the pressure of an artificial limb. Dedlow was transferred to the u.s. Army Hospital for Injuries and Diseases of the Nervous System

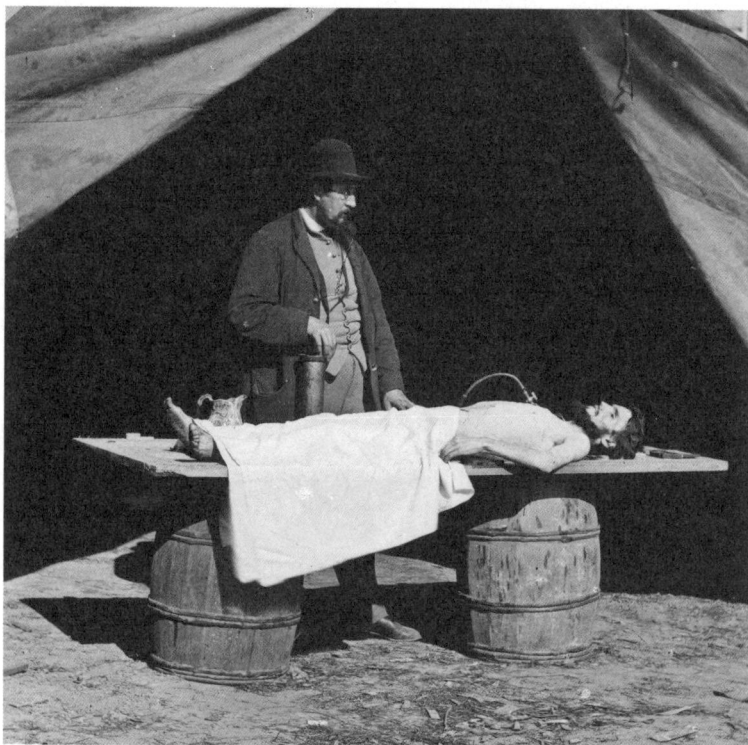

Dr Richard Burr pumping embalming fluid into the body of a dead soldier so that it can be transported to the deceased man's family, photograph *c*. 1863–5. Operations like the removal of bullets and amputation were carried out on the battlefield, on improvised tables.

in Philadelphia, commonly known as the 'Stump Hospital', since hundreds of its patients were being nursed back to health after the amputation of an arm, leg, both arms or both legs. Dedlow was the only one to have lost all his limbs. Every morning, he was carried in an armchair to the library. There was always someone who would read to him or fill his pipe. He had all the time in the world to think about his experiences as an amputee, and those of the men around him. George Dedlow – body shattered, but doctorly instincts intact – decided to draw up a medical report. A willing volunteer took his dictation.

He had noticed that, months after amputation, most men still had some feeling in their missing limb, in the form of an itch, pain or cramp. While those sensations persisted, it felt as if the limb was still there. And that made sense, Dedlow thought. Every stimulus to an arm or leg was transmitted by nerves to the brain. Even if nerves were severed at your ankle, knee or elbow, he reasoned, the bits that were left could still be stimulated, making it feel as if the hand or foot were still there. The nerve was like the bell cord used to summon servants from, say, the drawing room: even if someone pulled it in the middle, the servant would still think they were being called to the drawing room.

Tugs on the bell cord can be sparked as the wound heals; once that process is complete, the feeling that the limb is still there will ebb away. But sometimes the nerve damage is chronic, and the patient is continually reminded of what they have lost. The illusion is often detailed: an arm that has been amputated at the shoulder usually feels as if it is bent at the elbow. Sometimes the fingers are cramped so tightly into a fist that the patient feels the nails digging into their palm. In Dedlow's case, the pain in his left hand was sometimes so acute that in the night, groggy with sleep, he would grope with one missing hand for the other.

Another symptom related to Dedlow by patients who had been at the Stump Hospital for some time was the curious shortening of the amputated limb: a foot would feel as if it were attached to the knee, a hand as if it were at the elbow or even at the shoulder. Over time, a foot would creep ever closer: starting at its familiar spot, then moving up to the knee and finally the hip.

But Dedlow's missing limbs conjured up another sensation, more oppressive than any pain or itch could be. To his horror, he noticed that he seemed to be becoming less conscious of himself, of his own existence. At times this feeling of losing his identity, of no longer being the person he had been, became overpowering. He had to restrain himself from asking others whether he really was still George Dedlow. He kept his thoughts to himself, aware that they would sound absurd. He did have an explanation for them, though. Now that he had lost around half of the sensitive surface of his skin, his brain was receiving far fewer stimuli to process. Entire regions of his brain, previously in contact with the outside world, were now unemployed, had perhaps wasted away. Half of me, he reflected, is absent or inactive. A man is not his brain, he concluded, but the sum of his actions and perceptions. If part of those disappear, his sense of individual existence will also diminish. The thought weighed on his mind, and he fell into a depression.

The phantoms of Silas Weir Mitchell

George Dedlow's story was published in July 1866 in the *Atlantic Monthly*.[1] It was written in the first person, an autobiographical case study of the kind frequently published by doctors. The response was overwhelming. Flowers and letters of support for Dedlow were delivered to the Stump Hospital, and money was collected to make his life as a torso as bearable as possible. The *Atlantic Monthly* sent a cheque for $85 – back then a princely sum – to the author. Who was not George Dedlow.

Among the staff of that same Stump Hospital where Dedlow had allegedly dictated his story was a neurologist with literary aspirations: Silas Weir Mitchell. Like Dedlow, he was the son of a doctor and had trained at Jefferson Medical College. Nearly half a century later, in a lecture given to colleagues in 1913, the year before his death, Mitchell would explain what had happened.[2]

He had written the story about Dedlow for his own amusement. An acquaintance had read it and sent it to the *Atlantic Monthly* without his knowledge. Dedlow was a fictitious character. Mitchell had once seen the name above a jeweller's, and he liked the association with 'dead below'.

Amid the chaos of the American Civil War, medical statistics were recorded with astonishing precision. Three out of four surgical procedures were amputations. In total, some 60,000 limbs were cut off.[3] The operation seldom lasted longer than five minutes; an experienced doctor could do it in two. Sometimes a person lost both arms or both legs. But there are no cases on record of quadruple or even triple amputations.[4] The grim mortality statistics show why. Fewer than half survived amputation of both legs. Fewer than four in ten patients survived the amputation of both arms. The loss of Dedlow's one remaining limb – for the modern reader an almost Monty Python-ish twist – was a detail that Mitchell hoped would alert the reader to this being a work of fiction.

But perhaps he had by then mixed just a little too much truth into the story. It was a fact that the Union Army blocked the transit of medicines and anaesthetics. It was all too true that the type of munitions used by both sides – hollow balls of soft lead that flattened on impact, then tore a destructive path through the body – often made amputation unavoidable.[5] And above all, Mitchell had got his accounts about the curious sensations in missing body parts from men who actually had lost arms or legs. No wonder that his story was regarded as an authentic medical paper.

Mitchell was by no means the first to write about sensations in missing limbs.[6] The very first, in 1552, was Ambroise Paré. A Frenchman, he worked as an army doctor at a time when France was embroiled in one war after another. Paré applied a tourniquet just above the amputation site and devised new bandaging methods to stem the flow of blood. His innovations improved the survival rate for amputations. As a result, more men were able to report feeling pain in limbs that had been cut off months earlier. People who hadn't heard this with their own ears must find this very hard to believe, Paré admitted. But cases continued to be reported.

In a letter in 1637, René Descartes wrote that he had known a girl whose arm had had to be amputated because of gangrene. Before the surgery took place, she was blindfolded. The wound was so thickly bandaged that it was several weeks before she realized that her arm was no longer there. Descartes attributed

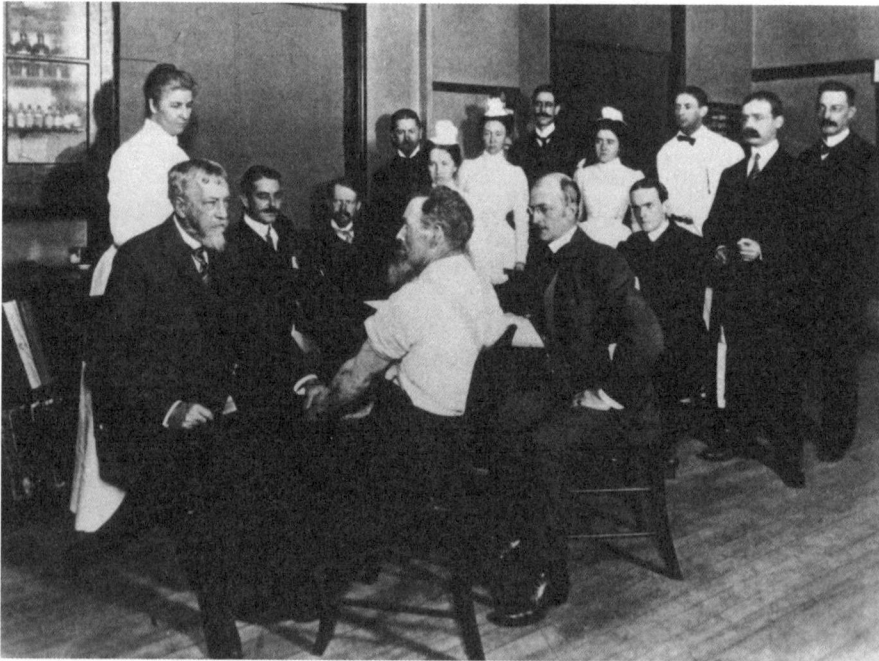

Dr Silas Weir Mitchell examining a Civil War veteran at his clinic
in the Infirmary for Nervous Diseases, Philadelphia, 1902.

the pain she still felt in the missing limb to the irritation of the remaining nerves. So the theory of the bell cord didn't originate with Mitchell – it goes back many centuries.

'Oh, the hand, the hand!'

In a classic investigative survey of London's poorest workers, first published in newspaper instalments in 1851 and 1852, the journalist Henry Mayhew quoted from an interview with a 'Negro crossing-sweeper who had lost both legs'. The man in question, Edward Albert of Kingston, Jamaica, had been a sailor whose legs became frost-bitten, and (after a disastrous attempt to thaw them out in a hot oven, where they got burnt) had to have them amputated just below the knee. During that interview, he said: 'Oh, yes, I feel my feet still: it is just as if I had them sitting on the floor, now. I feel my toes moving, like as if I had 'em. I could count them, the whole ten, whenever I work my knees. I had a corn on one of my toes, and I can feel it still, particularly at the change of weather.'[7]

In the literature of the first half of the nineteenth century, there are dozens of reports of sensations in missing limbs. So how did Mitchell manage to link *his* name to this phenomenon?

This didn't happen until he wrote about it again in 1871, this time under his own name.[8] The title of the article was 'Phantom Limbs'. His observations scarcely differed from those supposedly dictated by George Dedlow from his armchair in the library, but this time Mitchell wrote in the language of ghost stories, which were highly popular at the time. In almost every case, the lost leg or arm was still perceived as a spectre or phantom of that limb, he wrote, making his readers' flesh creep. 'There is something almost tragical, something ghastly, in the notion of these thousands of spirit limbs haunting as many good soldiers, and every now and then tormenting them with the disappointments which arise when, memory being off guard for a moment, the keen sense of the limb's presence betrays the man into some effort, the failure of which of a sudden reminds him of his loss.'[9] Mitchell would later write real ghost stories – now all forgotten – but the 'phantom' would remain.

He helped to perpetuate the term in publications for peers. In his handbook on nerve injuries, he wrote: 'Nearly every man who loses a limb carries about him a constant or inconstant phantom of the missing member, a sensory ghost of that much of himself, and sometimes a most inconvenient presence, faintly felt at times, but ready to be called up to his perception by a blow, a touch, or a change of wind.'[10] Phantom sensations were absent in only four of the ninety cases that he had investigated, and they were reported by all who had had arms amputated.

The illusion of a missing limb's presence could be extremely persistent. As one of Mitchell's patients put it: 'Every morning I have to learn anew that my leg is enriching a Virginia wheatcrop or ornamenting some horrible museum.'[11] Besides being persistent, the illusion could be deceptively powerful. Another patient, while riding a horse, had tried 'to pick up his bridle with his lost hand, while he struck his horse with the other, and was reminded of his mistake by being thrown.'[12] Another had tried to pick up his fork at every meal for nearly a year. He became so emotionally disturbed by this that he sometimes even threw up. Jumping out of bed when still half asleep, impulsively getting up from a chair, trying reflexively to catch a falling glass – it often ended badly.

A renowned French doctor, Alexandre Guéniot, had already written about the subjective shortening of limbs, a phenomenon now known as telescoping, in 1861. His patients, too, felt their amputated hands or feet slink ever closer, until they seemed to be right at the end of the stump.[13] In Mitchell's experience, the use of a prosthesis caused the hand or foot to move back to its old position. He noted a curious psychological component: when a prosthesis wearer thought or talked about telescoping, it felt as if their foot were suddenly back at their knee. Patients whose phantom sensations had diminished over time also reported that a focus on the stump and the missing limb made these feelings come back.

Mitchell found that manipulation of a stump caused sensations in a phantom limb. He asked patients to place their stump in ice-cold water, instantly conjuring up the sensation of an icy arm or leg. If their stump was placed in a draught, they felt the draught on their foot. Many patients told him that when their toes or fingers started to twitch, it was a sign that the wind would shift to the east. According to an old wives' tale, when someone died, the last thing they saw would be etched on their retina. That was a fable, Mitchell wrote, but he did know of cases in which the phantom sensations exactly corresponded with what the patient had felt just before the operation. In 1863, C., an army veteran, had been shot in the left arm. His thumb had flexed inwards, making the nail dig into his palm. He'd tried repeatedly to prise the thumb loose, but to no avail. He had lain like that for six hours before his arm was amputated. Now, nine years later, he could still feel the nail cutting into his palm. When a storm threatened, the cramp became worse.

Using electrotherapy – 'faradizing' – Mitchell could conjure up phantom limb sensations in patients who hadn't had them for some time. Once, without any warning, he had shocked the nerves of a man's shoulder: 'As the current affected the brachial plexus of nerves, he suddenly cried aloud: "Oh, the hand, the hand!" and attempted to seize the missing member. The phantom I had conjured up swiftly disappeared, but no spirit could have more amazed the man, so real did it seem."[14] This is almost the language of a spiritualist seance, with the doctor in the role of medium. Using his advanced apparatus, he could temporarily raise the dead.

The electric shocks to the stump corresponded to a tug half-way along the bell cord, creating the illusion that the foot or hand itself was being stimulated. In 1872 Mitchell had not got much further than Descartes in 1637 when it came to explaining phantom sensations. But his publications marked the beginning of a long series of studies – and frustrations – about their findings.

Buried eight times

Not long after the Civil War, American streets teemed with the many veterans – more than 45,000 – who had survived an amputation. They got around in invalid carriages, hopped on crutches or limped on prosthetic limbs. Between 1861 and 1873, more than one hundred patents were filed for prosthetics. In 1866 the state of Mississippi spent one-fifth of its budget on the purchase of artificial limbs.[15] Manufacturers of prosthetics, such as A. A. Marks of New York and Fisk & Arnold of Boston, both established in the year that the Civil War ended, advertised artificial limbs and rubber hands and feet. The artificial legs produced by Marks were made of willow and covered with parchment painted in flesh colour. In Boston, you could have yourself measured for an artificial leg made by the firm Douglass, with articulated knee, ankle and toe joints that – if their advertising lived up to its claims – could move noise-lessly and combined 'all the grace and beauty of the natural limb'.

A prominent citizen of Boston realized that this state of affairs presented a golden opportunity to study phantom sensations.[16] A good twenty years after the story about George Dedlow, the philosopher and psychologist William James wrote to the main prosthetics manufacturers in the region, asking for the addresses of their clients. They were only too happy to oblige the Harvard professor. A questionnaire was sent to around eight hundred people who had undergone an amputation. James received responses from 185 of them.

In his report, he stated right off the bat that he hadn't been able to establish why some amputees had phantom sensations and some didn't, why the position of the phantom limb remained fixed in some cases and roamed about in others, or why some could move their phantom limbs at will, while others appeared to have lost even the will to move them. Phantom sensations occurred whether a stump was still painful or had healed well. In some cases the illusion wore off; in others it didn't. A man in his seventies told James that his leg had been amputated when he was thirteen, yet he felt his phantom foot as vividly as his existing foot.[17] Two in three could wiggle the toes of their phantom foot. This nearly always involved muscle movements in the stump – though even if those muscles were no longer there because the leg had been taken off at the hip, some could still move their phantom toes. By contrast, others still had the necessary muscles, but their phantom toes felt completely immovable. Especially in the beginning, the illusion that a missing body part was still there could be overwhelming: a person with an amputated arm would reflexively grab a pair of nail scissors to cut their invisible nails. James, like others who'd studied phantom pain, found that it sometimes mirrored what a patient had felt just before an amputation: one of his respondents was still troubled by a blister on the heel of his phantom foot; another still felt the pain of his chilblains.

James, too, thought that phantom sensations were probably caused by stimulating the remaining nerves. If so, the real mystery was why some people did *not* have such sensations. Perhaps in their case, the nerve ends were so embedded in the stump that they could no longer be reached by stimuli. He could think of no other explanation.

James published his study in 1887 in the journal of the *American Society for Psychical Research,* an organization that he himself

The first recorded amputee of the American Civil War was the eighteen-year-old engineering student James Edward Hanger, who had volunteered with a Southern cavalry regiment. On 3 June 1861 he was hit by a cannonball, and his left leg had to be amputated about 18 centimetres (7 in.) below the hip. After recovering, he designed an artificial leg with hinges at the knee and ankle. The company he founded is to this day a leading manufacturer of prostheses. In this photograph, c. 1901, five men with a 'Hanger limb' pose in front of one of its branches.

The stand of the firm A. A. Marks at the World's Columbian Exposition in Chicago in 1893 offered free brochures in English, French, Spanish and German.

co-founded a few years earlier. Perhaps to avoid misunderstandings, he did not talk about phantoms, spirits or ghosts of limbs, but simply about 'lost limbs'. He did, however, add a few final remarks that seem specially aimed at readers with an interest in the supernatural. It was said, James wrote, that shortly after an amputation, some people could still feel what was happening to the limb that had been cut off. One veteran, for example, felt a painful gnawing sensation in his amputated leg. This prompted his friends to dig up the leg: they found maggots crawling in it. After the leg had been burnt, the pain disappeared. There were many versions of this story. Around twenty people had written to James about similar phenomena. Most accounts were vague. James felt there was a rational explanation: pain in the first few weeks after amputation takes different forms, and there would always be a moment when the phantom sensations seemed to correspond with the amputated limb being buried, burnt or plunged into formaldehyde. It has to be said that his correspondents included a few oddballs: one man wrote that he had already dug up his amputated leg eight times in order to reposition it. He wanted to know if he should continue to do this. He hoped not.

The elusiveness of phantoms

Reading between the lines, you sense William James's irritation about the lack of clear conclusions. And the picture is no different today. Research into phantom sensations is still characterized by differences between individuals and differing symptoms in one and the same patient, of rules with too many exceptions and of therapies that work for some and not for others with precisely the same injuries.

A century and a half after Mitchell's first publications, it is no longer limbs torn off by cannonballs that come to haunt their owners. These days, when legs have to be amputated, more often than not the reason is damage to blood vessels, usually from diabetes. By contrast, in most cases (75 per cent), arms are amputated because of accidents. In war situations, the most common cause of amputation is injury due to land mines or roadside bombs.

Hundreds of studies into phantom sensations have meanwhile shed more light on phantom pain, yet also deepened the mystery.[18]

The fact that phantom pain is most severe in the first days and weeks after the operation seems logical, but why does it peak again a year later? Is that a 'last-year-at-this-time' effect? Why are women more prone to phantom pain than men? Why are phantom sensations more intense during an orgasm for women, yet only afterwards for men? What *has* been established is that phantom pain is more frequent after the amputation of arms than legs, and that it makes no difference whether the amputated limb was on the left or right. On the whole, phantom sensations diminish in frequency and intensity over the years. However, for around 10 per cent of amputees, phantom pain becomes chronic. An increase in pain is rare. Phantom sensations mainly occur at times of rest, during the evening and night, and can cause sleep disturbances. Stress or cold makes them worse.

In one in five cases, telescoping occurs. This is less likely if there was no previous nerve damage, for example when an amputation is done suddenly, after an accident. When a prosthesis is used, the phantom hand or foot usually returns to its old spot. Using a prosthesis early on does not reduce phantom sensations, though it does reduce pain.

Phantom sensations need not necessarily take the form of pain or an itch or cramp. A person can, say, have the feeling that their lost arm is sticking out to the side. The illusion is so convincing that they walk sideways through doorways, so as not to bang it. Others feel as if they're holding a cup in their hand. In patients with heart problems, the pain can radiate out to the lost left arm. Existing pain, for example from an ingrowing toenail, can continue after amputation and morph into chronic phantom pain. In 1797 Horatio Nelson was shot in his right arm at the Battle of Tenerife and for the rest of his life felt his fingers digging into the palm of his hand – just as they had in the hours before his arm was taken off by the ship's surgeon.

Sensory phantoms

Not all phantom sensations relate to touch. If a malignant tumour develops in an eye, the whole eye has to be removed. In a study of 179 people who had lost an eye because of a melanoma, half said that they still received visual input from the lost eye.[19] One in every four patients occasionally felt that they were able to see with the lost eye. A man said he'd seen someone walking next to him, while a woman reported seeing an unknown figure by her bed. Just like phantom sensations from amputated limbs, these visions were most common when it was dark, or when the person was tired or resting. They weren't caused by the remaining eye, because visions were also experienced by patients who'd lost both eyes, usually after accidents with explosives.[20] Indeed, it's considered good practice to warn patients before the operation that they might have visual hallucinations.

Some people who have lost their hearing after damage to both ears, or injuries to the parts of the brain that process acoustic stimuli, still hear phantom sounds. These are even spookier than phantom touch sensations: it's as if they hear someone talking in a room they know to be empty. The sound is just there, a split second before the patient remembers that they *can't* hear any more.

Phantom smells are equally unsettling. Take the case of a 25-year-old man who had suffered severe brain damage in a car accident.[21] The areas that process smell stimuli had been damaged on both sides of his brain, so he lost the ability to smell anything – even really pungent odours like ammoniac. A few months after the accident, though, he began to experience phantom smells. Some were pleasant, like the perfume of roses or cinnamon, others less so, like vinegar or the smell of something burning. The phantom smells occurred a few times a month, and three years after the accident had not diminished at all. They were fleeting, lasting only a few seconds, but powerful all the same: the man scored their intensity as eight out of ten. These wafts of smell were so realistic that he often checked just to see whether something wasn't burning on the stove.

Phantom genitals

There is one last category of phantom sensations that only doctors would think – or have the nerve – to ask about. Can one still have sensations in a lost penis? Or an amputated breast? The answer is yes.

In 1786 a Scottish doctor called John Hunter wrote: 'A serjeant [sic] of marines who had lost the glans, and the greater body of the penis, upon being asked, if he ever felt those sensations which are peculiar to the glans, declared, that upon rubbing the end of the stump, it gave him exactly the sensation which friction upon the glans produced, and was followed by an emission of the semen.'[22] A colleague of Hunter's wrote about a man by the name of W. Scott, whose penis had been almost entirely shot off. The stump was so small it hardly projected, but it remained sensitive to sensations that seemed to come from the glans.[23] Much more recently, in 1950, a surgeon reported the case of a man whose penis had been so damaged in an accident that amputation was unavoidable. Every now and then he felt phantom erections. He could not bring them about himself, nor were they caused by sexual arousal, but the illusion was so strong that he sometimes checked to see if his penis had really gone.[24] In 1999 the neurologist Charles Miller Fisher described the case of a 64-year-old whose penis had been amputated twenty years previously due to cancer. He, too, had phantom erections, but in his case they were caused by erotic stimuli.[25] In a group of twelve men whose penises had had to be amputated due to cancer, six reported phantom erections or other phantom sensations.[26] None of the men experienced the erections as painful.

Even a penis that is voluntarily relinquished can live on in phantom form. Transgender people have the feeling – often from a very early age – that they are trapped in the wrong body. As a result, they may opt for hormone therapy and surgery. In male-to-female gender reassignment surgery, the penis is usually removed. The neurologist Vilayanur Ramachandran and his colleague Paul McGeoch – when both researching cognitive neuroscience at the University of California, San Diego – raised a few intriguing questions.[27] Someone assigned male at birth but who knows herself to be a woman will perceive her penis as 'alien' to her subjective identity as female. Might trans women therefore be less likely to experience a phantom penis after amputation? That hunch proved correct. Out of a group of twenty trans women, only six (30 per cent) experienced phantom sensations after the operation. In the case of men who had undergone the penis amputation for different reasons, such as cancer, twice as many (60 per cent) did so.

Two years after a mastectomy, almost one in five women still experienced a phantom breast.[28] In the case of a group of 29 trans men who had undergone gender-affirming surgery – involving the removal of nearly all breast tissue – this percentage was much lower: only three of them reported still occasionally feeling a phantom breast. But the transgender group studied by McGeoch and Ramachandran produced a yet more surprising finding. Eighteen of them (62 per cent) said that even before the operation, they had occasionally experienced phantom penises and phantom erections – sensations that went back many years, sometimes as far back as childhood. These were actual sensory experiences, not fantasized or imagined penises. It even felt as if the penises had a specific length or curvature. Several trans men experienced morning erections or erections while asleep. They learned from the researchers that men indeed have erections around five times a night during REM sleep. A control group of cisgender women were also asked if they had ever had the sensation of a phantom penis. None had.

It's all in your mind

Right from the start, phantom sensations have prompted questions that branched out from surgery to such disciplines as neurology, psychology, psychiatry and even philosophy. After an amputation, where do the borders of your body lie? Does your sense of self include your lost body part? Or is it only present when it seems to be generating sensations? Is your body what you feel, or what you see in the mirror? What does it mean that three out of ten people with an amputated leg can still walk in their dreams, even years after the amputation?[29] Or that someone who went blind at the age of seven continues to dream in images for the rest of their life?[30] Can a body part really be missing if the body persists in remembering it, as if unable to resign itself to the loss?

Another notion, the nerve as a bell cord – popular well into Victorian times – has also bitten the dust. Over the course of half a century, Mitchell's hypothesis that phantom sensations were caused by the stimulation of nerve endings in the stump (now known as the peripheral theory) was gradually dismantled. What goes on in the stump is not entirely separate from the occurrence of phantom sensations, but it is only part of the story – and not the most important part. The arguments against the peripheral theory are short and convincing. When a stump is locally anaesthetized – meaning that no tug on the bell cord can come from that spot – people still experience phantom sensations. Severing a nerve just above the scar tissue does not end phantom sensations. Patients who experience such sensations in an amputated foot sometimes have their lower leg amputated in a later operation. Although this removes the old nerve endings, some still experience phantom pain that appears to come from both the lower leg *and* the foot.

Finally, people who are born without arms or other parts of limbs can also experience phantom sensations, despite not having damaged nerve endings. They might have cramp in a leg that was never there, or gesticulate with non-existent hands. Just as the bell cord has today vanished from most households, Mitchell's hypothesis has made way for the central theory: phantom sensations arise not in the nerve endings but in the brain. They are not caused by the stimulation of nerve endings but rather by a *lack* of stimulation, prompting activity in the brain. These days – apart from in period dramas – bell cords only live on in the popular myth about phantom sensations because the idea that they're caused by scarring or stump stimulation is extremely hard to shake. The fact that such sensations are now thought to have a neurological cause (literally 'all in your mind') is the result of a strange mix of discoveries encompassing epilepsy research, macaques and mirrors.

Homunculus

The short life of Mary Rafferty began as miserably as it would end. As a baby, she fell into a fire. Her scalp was so badly burnt that hair would never grow on it again. Under the wig that she now wore, thirty years later, a fast-growing sore had appeared. In early 1874 she was admitted to the Good Samaritan Hospital in Cincinnati, where she fell into the clutches of Roberts Bartholow, professor of clinical medicine. In April that year, he would personally record everything that happened to Mary after she had come to him, innocently hoping to be helped.[31]

Mary, an Irishwoman who worked as a domestic servant, struck Bartholow as having a sunny disposition. 'She is cheerful in manner, and smiles easily and frequently,' but for the rest he dismissed her as 'rather feeble-minded'.[32] The cancerous sore was exactly centred above the two halves of her brain and had eaten a hole in her skull 5 centimetres (2 in.) across. Peering into it, Bartholow could see the brain pulsating. Her exposed brain gave him a perfect opportunity to do some tests. He led her to his 'Electrical Room', a laboratory for electrotherapy and experiments that he'd had installed a few years earlier.

To Mary, the surroundings must have seemed intimidating. The room was packed with machines to generate and store galvanic (direct) and faradic (alternating) electrical currents, some the size of a large sideboard. Bartholow instructed her to take a seat, while he selected two electrodes, connected to a faradic battery. He then bent over her head and started poking them into her brain, without switching on the current. She did not react at all, proving – as he had (rightly) assumed – that the brain itself is insensitive. But when the electrodes were charged with the weakest possible current, pricking the left side of the brain did have an effect: Mary's right arm and right leg shot up. She complained about an unpleasant tingling feeling and began to rub her right arm hard with her left hand. For Bartholow this wasn't enough; he had hoped for 'more decided reactions.'[33] He now stuck the electrodes in the right side of the brain and ramped up the current. Given his detailed observations of the consequences, he must have done this for some considerable length of time.

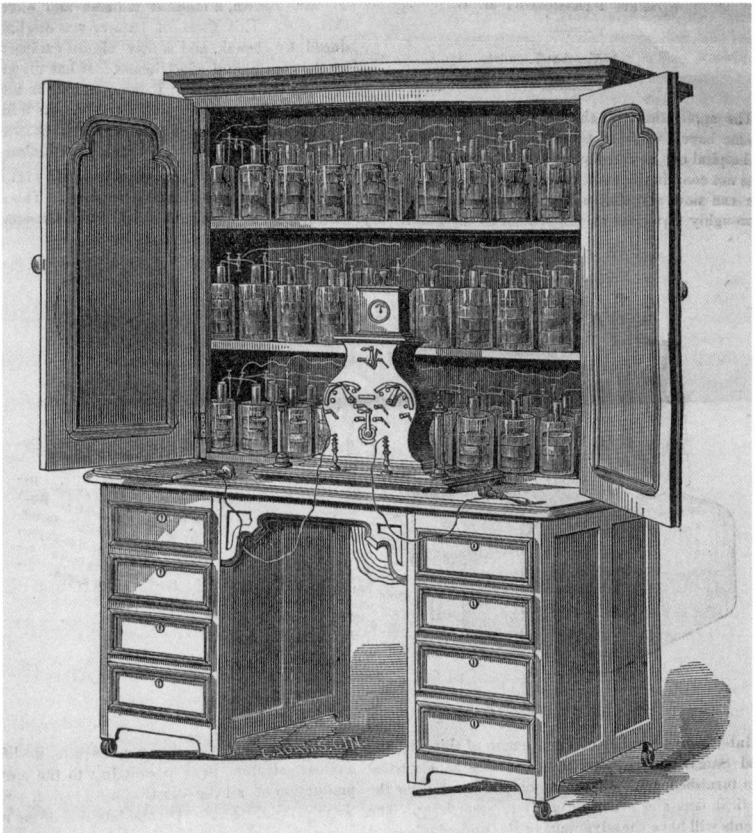

The fearsome 60-cell galvanic battery manufactured by Siemens & Halske. By the time Bartholow was ready to start performing galvanic experiments on Mary Rafferty's brain, she had already been so weakened by the ones with faradic current that he had to abandon further tests.

Mary immediately began to cry. Her left arm spasmed. She stared unseeingly, with greatly enlarged pupils. Her lips turned blue. She foamed at the mouth. Her breathing became laboured, and she lost consciousness, lapsing into a coma. Twenty minutes later she regained consciousness, feeling weak and dizzy. The electric shocks induced by Bartholow had sparked an epileptic fit. She would never recover.

A few days later, Bartholow was keen to continue his experiments, this time with galvanic current. But when Mary was once again brought to the Electrical Room, she was in such a wretched state, her entire body wracked by spasms and cramps, that he decided against it. Just before she threatened to slip into another coma, an assistant anaesthetized her with chloroform. Another severe epileptic fit followed, paralysing her completely on the right-hand side. Then she died.

The cheerful maid who a week earlier had come to Bartholow merely to be cured of a sore would go down in history as the first patient whose brain surface had been electrically stimulated for experimental reasons. Bartholow alleged that her death was caused by an infection of the arachnoid or 'spiderweb' membrane, the diaphanous tissue protecting the brain. But the autopsy showed that two electrodes had been driven deep into her brain – one to a depth of 4 centimetres (1½ in.).

In 1937 Mary Rafferty featured in the introduction of a now classic article written by two Canadian neurosurgeons, Wilder Penfield and Edwin Boldrey.[34] In 1928 Penfield had devised a surgical technique for the diagnosis and treatment of patients with severe epilepsy. It involved probing the brain with electrodes – but with a very weak, carefully calculated current. First, EEG scans were made of the shaved skull to determine the site of the epilepsy.

Then, under local anaesthetic, part of the skull would be removed, exposing the surface of the suspected area. Using an electrode that emitted a minuscule, five-millisecond electric charge, the surgeon would probe the brain very gradually, fold by fold. During this process, the patient would remain conscious. That was crucial, because they needed to report their sensations. Depending on the spot that was touched, these might be sounds, flashes or a tingling feeling; an arm might suddenly move; the patient might feel that water was dripping on their hand, or might briefly lose the power of speech. If a touch led to an epileptic fit (or an 'aura' heralding a fit), then the real operation could begin: the patient would be fully anaesthetized and the section of brain tissue in question removed. Penfield did not shrink to use his scalpel: he was of the school 'no brain is better than bad brain'. He would mark areas that prompted a specific reaction when touched by placing a small, numbered scrap of paper on them. Then he would have the surface photographed. In this way, he gradually built up an advanced map of brain functions. By 1937, 163 operations later, he could indicate exactly where sensory stimuli were processed or the motor functions of hands, legs or lips were controlled. For tongue movements alone, he identified sixteen areas. Even the movements of each of the five fingers turned out to be controlled from four or five different locations.

One of Penfield's later diagrams featured an image of the brain's surface on which he drew sensory and motor 'homunculi', tiny human figures whose proportions reflect the size of the brain regions devoted to individual body parts. Both homunculi look grotesquely deformed. The brain's arm region is much smaller than the hand region, so the motor homunculus has skinny little arms and whopping great hands. The brain's focus on sensitivity

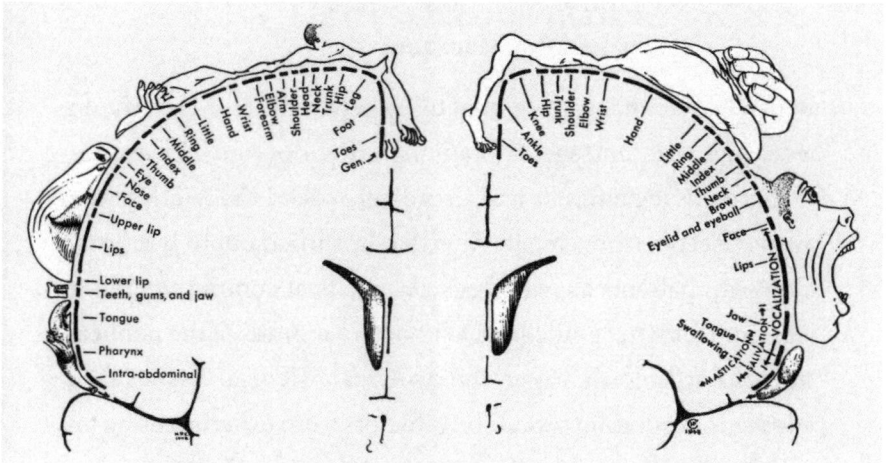

The sensory (left) and motor (right) homunculus of a man (Penfield did not devise a female equivalent). For the sake of clarity, he positioned them on either side of the brain. In reality, each side of the brain has both a motor and a sensory homunculus.

in fingers, tongue and lips makes the sensory homunculus look equally weird: it consists mainly of hands and face.

Looking at Penfield's brain maps, you can see at a glance which areas control the movements of an index finger, thumb, tongue or eye. Those positions are hardwired in our brain and are largely universal. It was long thought that, after childhood, this was a fixed, unalterable state of affairs. In a young brain, damage to Broca's area – linked to speech function – could still be compensated through relocation to the other side of the brain, but in the case of adults, the functions and their location were set. Back in 1866 Silas Weir Mitchell had no idea which parts of the brain all those bell cords were attached to. By the 1950s, thanks to the work of neurotopographers like Penfield, it was apparently clear which areas would fall into disuse after, say, the amputation of a right index finger.

Macaques

Bartholow's chilling account of his experiments on Mary's brain were met with outrage, also among other doctors. His experiments on living animals had already provoked the wrath of the antivivisection movement. Now that he turned out to be as cavalier with patients as with dogs, the medical community felt it was time for strict guidelines. Within two months of the publication of Bartholow's paper, the American Medical Association adopted a resolution prohibiting doctors from experimenting on their patients, 'except for the purpose and with the hope of saving said patient's life, or the life of the child *in uterus*.'[35]

The same protection was not given to laboratory animals. Even primates were considered fair game. In the history of research into phantom sensations, it seems as if not a single step could be taken without people or animals having to suffer. In 1981, in Silver Spring, Maryland, seventeen macaques were seized from the laboratory of the psychologist Edward Taub.[36] A few years earlier, the nerves that transmitted sensations from the arm to the brain had been severed near the spine in the then four-year-old animals. As a result, their arms were insensitive to touch and pain stimuli. The monkeys could no longer feel heat or cold, or the position that their arms were in. In some animals, the nerves of both arms had been severed. For monkeys that washed or rubbed their food clean and used stones to crack nuts, these were dangerous and disabling injuries. Taub hoped through his research to discover tools for the rehabilitation of people whose limbs had become paralysed after a stroke, or as the result of congenital brain injury.

The seized monkeys were placed in the buildings of the National Institutes of Health, where for years they pined away in tiny cages. In 1987 five monkeys that could still just about function were transferred to the San Diego Zoo. The rest were too severely disabled and ended up in a primate centre. Vets there advised that four of the monkeys be euthanized. Two had mutilated their own arms, probably because of phantom pain; a third monkey developed gangrene and had to have its arm amputated.

But a humane death was not yet a prospect for this battered quartet. The neuroscientist Timothy Pons requested permission to subject their brains to a last series of tests – seeing as they were going to be killed soon anyway. Permission was granted. Twelve years after their nerves had been surgically severed, the monkeys – by now quite elderly – were back in the laboratory. That these experiments would yield the first convincing explanation for phantom sensations contrasts uncomfortably with their hopeless suffering.

The monkeys were pinioned in vices inside a metal frame. The portion of skull above the brain area responsible for the insensitive arm was removed, and a mechanically operated electrode methodically probed the brain surface. Given that the area had received no input for the previous twelve years, Pons didn't expect to receive any more signals from it. To his surprise, much of it had been taken over by an adjacent area – the region that represents the face in both monkeys and humans.[37] Very light touches to the faces of the monkeys – it was enough to just push a few hairs aside – activated not only the normal brain map of the face but also part of the arm map. The brains of the macaques turned out to have *two* maps of their faces.

Ten years prior to these experiments, it had already been demonstrated that, after an amputation, the outer perimeters of corresponding brain areas were activated by adjoining areas. The metaphors used were of invasion and colonization: areas that had been freed up were being claimed and occupied by neighbouring areas. The thinking was that the invasion would be restricted to 1 or 2 millimetres from the edge, since axons – the fibres that project from a nerve cell – wouldn't be able to reach further. As an explanation, it left intact the image of a hardwired brain, with a static mosaic of functions and locations. Pons disproved this theory. He found that the new face map spanned the entire arm map, extending some 10 to 14 millimetres (¼–½ in.) across the cortex. Such activation couldn't possibly be caused by the axons on the perimeter. This was no cautious incursion of the border region; it was annexation.

Now, at last, the monkeys were euthanized. The territorial changes that Pons had seen in their brains sparked a revolution in neurology. The old image of the brain as a collection of circuits – a system in which each severed connection would remain dead – made way for a concept that featured plasticity and adaptation. Even in an adult brain, it seemed, new connections could be made. Borders blurred. Circuits did not shut down; they were assigned different tasks. After amputation, the old cortical maps were no longer reliable.

Mirrors

More than a century after William James's questionnaire, various prosthetic manufacturers – this time in the San Diego region – received a similar appeal from a professor. The neuroscientist Vilayanur Ramachandran was looking for volunteers for what seemed like the next logical step after Pons's findings: investigating whether humans also formed new cortical maps after an amputation.

Seventeen-year-old Tom had lost his left forearm four weeks earlier in a car accident.[38] A Ramachandran-esque detail: as Tom flew through the air, he saw his own hand still gripping the chair seat. His arm was amputated 6 centimetres (2 in.) above the elbow. Phantom sensations soon followed. Whenever he stumbled, the arm would shoot out to break the fall. The phantom hand often itched. At their meeting in his office, Ramachandran asked Tom to sit down and close his eyes. He began to methodically touch the surface of Tom's skin with a cotton bud, giving each spot two quick, light strokes. Tom was asked to say where he felt the touch. When Ramachandran touched his lower jaw, Tom said he felt the touch in his cheek but also in his phantom arm. A touch on his upper lip was also felt in his phantom index finger. After a long, painstaking exploration, Ramachandran could, by touching Tom's face with his cotton bud, conjure up sensations in the phantom hand finger by finger. When warm water was applied to his cheek, he felt the heat in his phantom hand. When it trickled down his cheek, he felt it running down his missing hand. Touch sensations in the left hand could now apparently be generated from a map on the face. A touch in the middle of his face felt to Tom like a touch on his thumb, whereas a touch slightly to the side felt like a touch on his pinkie. He could finally get rid of the itch in his phantom hand by rubbing a spot on his cheek. About 7 centimetres (3 in.) above the amputation line, Ramachandran found a second map for the hand.

A week later, the new maps were still located at exactly the same spot. Experiments with a second test subject, K., who had lost his entire right arm a year earlier, confirmed the findings in Tom's case. K., too, had a new map in his face, with the phantom hand on his lower jaw and elbow at the corner of his jaw. Just as with Tom, a second map was found in K.'s case, close to the line of amputation, just under the armpit. As Pons had already noted in his tests with the monkeys, a very slight touch was enough: Ramachandran only had to brush a single chest hair for K. to feel it in his phantom elbow.

The maps matched phantom sensations in very specific spots with astonishing precision. Could this be the result of axons growing fibres to regenerate body parts deprived of nerve supply (axonal sprouting)? It seemed unlikely – for that, the four weeks between amputation and the experiment with Tom seemed too short. Perhaps, Ramachandran surmised, the maps weren't new at all; they had always been there, just dormant, and their activity had been suppressed as long as the brain still received stimuli from a healthy limb. But whatever the explanation, the brain was clearly more plastic than previously assumed.

Ramachandran's findings prompted a new theory about phantom sensations. It seemed that, after an amputation, the brain area that had formerly received input from the lost body part did not just sit there inert and disconnected. On the contrary: the very fact that it had been cut off made it vulnerable to infiltration, invasion or even complete occupation by a neighbouring area. The resultant 'foreign' activity was felt in the missing body parts. Phantom sensations did not originate in severed nerve endings but in the brain's sensory cortex.

All kinds of curious observations and experiences now fell into place. One man had felt sensations in his phantom leg when shaving and dismissed this as merely bizarre. Ramachandran was able to explain that it was actually very logical. There was also now an elegant explanation for telescoping: the brain maps of the upper and lower arms – already comparatively small – shrink after an amputation and soon generate far fewer phantom sensations than the much larger map of the hand. This makes patients feel as if their arm has disappeared and their hand is now attached directly to their shoulder. Ramachandran also hazarded a prediction. On Penfield's sensory homunculus, the neural maps of the foot and toes lie next to the genital maps. After a penis amputation, stimuli to the foot should conjure up sensations in the phantom penis. The reverse should also be true. Indeed, Ramachandran subsequently heard from a number of men who had had a leg amputated that they now felt orgasms not only in their penises but in their phantom legs and feet.[39] Moreover, their orgasms were now more intense, perhaps because the freed-up foot and leg areas had been activated together with the area for the penis.

Observations like these were proof of neural plasticity: the brain's adaptability meant that a patient continued to have phantom sensations. But in the case of phantom pain, especially when it was chronic, this was an unwelcome kind of plasticity, one that turned against their own body. A common type of phantom pain among those who came to see Ramachandran was a feeling of paralysis or painfully clenched fists.[40] Sometimes a person could unclench the phantom fist by mentally straightening each finger in turn, but that could take three-quarters of an hour. However, most couldn't get their phantom fingers to move at all. A paralysed phantom arm was equally unpleasant. The issue was not lack of sensation but rather that the arm felt dead and immobile, as if cast in concrete. The latter was true for the 28-year-old S., whose left arm had been amputated above the elbow ten years previously, after a year in a sling.

An arm that is no longer there can no longer show how a hand or finger responds to motor instructions. A patient might will their arm to move, but when nothing happens – when there's no visual confirmation that the arm is responding – the brain keeps repeating the instruction. It was precisely this impasse that Ramachandran overcame with a construction as simple as it was ingenious.

He placed a box on the table. It had two openings in the front and was divided in two by a mirror placed vertically in the middle. There was no top, so you could see inside. Ramachandran asked S. to put his phantom arm through the left-hand opening, behind the mirror, and his remaining arm through the right-hand opening, in front of the mirror. First, he had to move both arms as if conducting an orchestra, with his eyes shut. He could do so with the right arm, but the left one still felt frozen. He was then allowed to open his eyes and bend a little to the right, so that he could see both his right hand and its mirror image. Again he was instructed to move both arms. The effect was instantaneous and spectacular. Just like Silas Weir Mitchell's patient, who after an electric stimulus to the shoulder suddenly felt his missing hand again and cried out 'Oh, the hand, the hand!', so S., too, cried out in surprise: 'Oh, my God! Oh, my God, Doctor! This is unbelievable. It's mind-boggling. My arm is plugged in again; it's as if I am back in the past. All these years I have often tried several times a day without success, but now I can actually feel I'm moving my arm, Doctor.'[41]

Disappointingly, when Ramachandran removed the mirror or asked S. to perform the same movements with his eyes shut, the sense of paralysis returned. S. was given the box to take home and was asked to practise this exercise for 15 minutes every day. A week later, the trick still worked – but, without the mirror, his arm remained immobile. Three weeks later, S. rang up excitedly to say that his phantom arm had disappeared, leaving only a few fingers. They felt as if they were attached to his shoulder stump, and he could move them. Six months later, even though the box had not been used, the phantom arm had not returned. With a literary flair reminiscent of Mitchell's, Ramachandran wrote: 'What we had achieved, therefore, may be the first known case of an "amputation" of a phantom limb!'[42]

The visual feedback of a paralysed or missing arm appearing to move makes the brain 'think' that its commands are once again being followed, enabling tightly clenched fists to be opened, or an arm that has been painfully bent for years to be stretched at last. Ramachandran's therapeutic illusionism proved extraordinarily effective. Today variations of the mirror box can be found in

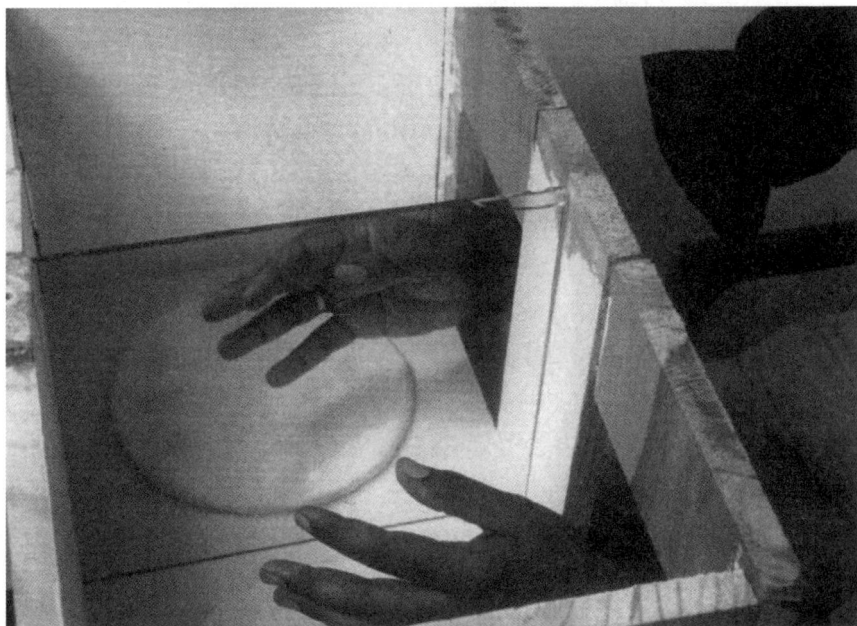

Ramachandran's 'mirror box'. Unlike in the case of S., here the functioning left hand is mirrored. Behind the mirror – invisible to the patient – is the stump of their right arm, or their paralysed right hand. Thanks to the mirror, movements of the left hand are seen and perceived as movements of the right hand.

rehabilitation institutions all over the world, not just for the treatment of phantom pain but more commonly to restore movement to limbs paralysed after a stroke. The technique has its limitations: there must be a functioning limb to provide the mirror image – George Dedlow would not have been helped by a mirror box – but thanks to advances in virtual reality, other solutions have been devised.

Neuro-illusions

The story about George Dedlow, Mitchell related, looking back on his years in the Stump Hospital, had a curious origin.[43] One evening, a friend who was visiting him had come up with an interesting question. How much would someone have to lose of their body before they lost their feeling of individuality? They'd debated it for some time. Mitchell had started writing his story that same evening. Dedlow was a thought experiment – the conclusion being that the young man's long, lonely hours as a torso in the library gradually eroded his self-awareness. Without arms and legs, he was only a fraction of his former self. It seems that Mitchell saw consciousness of identity as closely connected to what was still intact and present in a person's own body.

Might Mitchell have got this idea from conversations with the veterans in his hospital? It seems unlikely. After an amputation, people do not feel that the contours of their physical self-image shrink according to what has been lost. That image stays the same. It just no longer coincides with the contours of the body that they see in the mirror. The same applies even when limbs have been missing since birth. People born without arms don't perceive their physical self-image as ending at the shoulders. The brain areas that should have directed the arms are still intact, even though they have never received feedback of movement. People without arms feel phantom arms moving back and forth as they walk, or raise a phantom hand to hail an acquaintance. Just like everyone else, they have a preference when folding their arms: left over right, or vice versa.

When they tell people this, the response is often sceptical. Hands that have been missing since birth, and yet still gesticulate? It seems like a tall story. Aren't they just copying gestures they see others make? Or imagining how it must feel to move your hands? A simple but convincing test puts an end to such doubt. Try making a loose fist, and then raising your fingers and folding them back into your fist one by one: thumb, index finger, middle finger, and so on. It works with every finger – except one. The ring finger gets stuck halfway; most people can't even get it to an angle of 45 degrees. Almost no one knows that the ring finger is the odd one out. People who have been born without hands don't tend to know this either. But if they are asked to stick up each finger from their phantom fist, they experience the same problem when they try to stretch their ring fingers.[44]

Such distrust is telling. Back in 1552 the army doctor Ambroise Paré had already predicted that phantom sensations would meet with disbelief. They are traditionally described in terms like 'illusory', 'deceptive', 'imaginary' or 'apparent', as if there were nothing true or real about them. The term 'phantom' in particular suggests a kind of pseudo-reality. From there, it is only a small step to the suggestion that phantom sensations are psychosomatic, the product of an inability to deal with the loss of an arm or a leg. The thinking goes that patients are in denial, subconsciously rejecting this terrible breach of their physical integrity. However, hundreds of studies into the relationship between phantom perceptions and the neural maps of limbs (and other body parts) carried out in recent decades have since relegated this theory to the same dustbin of history as Mitchell's bell cord.

Up until the 1960s, lack of a convincing neurological explanation spawned hybrid theories: half neurological (stimulation of damaged nerve endings), half psychological (denial). These days, phantom sensations are a welcome window into the plasticity of the brain. They have helped to oust the belief that brain circuits, once in place, can never reorganize, as well as to clarify the distinction between the *body scheme* (the maps of the various body parts in the brain) and the *body image* (the subjective perception of what belongs to one's own body). Phantom sensations show that the contours of the two do not always coincide, that a lost leg, arm or penis can still be perceived as belonging to the self, even if the relevant brain map no longer receives any signals from the amputated body part.

A century and a half after George Dedlow's sombre musings, the question of whether a reduced body diminishes self-awareness no longer needs to be answered from an armchair. We now have access to observations, experiences, experimental findings and measurements of the relationship between body and body image. People who learn to use a prosthesis after an amputation find that its success depends on the extent to which they can make it their 'own' and incorporate it into their body images. These mechanical aids need to be animated if they are to function optimally. As long as they're seen as an alien extension to the body, phantom sensations often persist, whereas integration suppresses them. The feeling that a prosthesis is part of one's body can go so far that an unexpected blow to an artificial hand causes as much shock and pain as a blow to a living hand.

Versions of Ramachandran's mirror experiments have also been carried out on people with intact limbs, to see if their body image can be manipulated. This turned out to be surprisingly easy. In the 'rubber hand illusion', a person is asked to place their hands on a table, palms downwards. A screen prevents them from seeing their left hand. Instead they see a lifelike rubber hand positioned next to their right hand. The experimenter then strokes both the rubber hand and the hidden hand with two small brushes, at exactly the same time and in the same spot. After only a few minutes, the person starts to perceive the rubber hand as their own. The discovery of this illusion in 1998 inspired researchers to use it in other tests.[45] Halfway through one such experiment, Ramachandran bent one of the rubber fingers into an anatomically impossible position. The test subjects felt the pain as if it had been their real finger, a finding that was confirmed by skin conductance measurements.[46] The essence of all variants remained the same: the combined evidence of what the hand feels and the eye sees makes people believe that it's their own hand that's being stroked or injured. You don't have to have a limb cut off to feel the contours of your body image shift.

The integration of a prosthesis and the rubber hand illusion both depend on extending the contours of a body image. But what happens if contours are drawn too closely? After certain neurological injuries, like a tumour or a brain haemorrhage, body parts that have become paralysed can suddenly be perceived as 'alien'. An arm or leg no longer feels part of your body. It's a deeply disconcerting experience. The leg lying in your bed must have been switched with someone else's – perhaps even with a dead leg. You need to get rid of it as soon as possible. In more than one case, agitated patients have fallen out of bed while trying to shove their leg over its edge.[47] This syndrome is named after Otto Pötzl, the Viennese neurologist who first described it in 1935. Sufferers have a distorted body image that fails to include an arm or a leg, or even an entire half of the body – the opposite of what happens with the rubber hand.

Silas Weir Mitchell had in fact already encountered this phenomenon back in the nineteenth century. If anything, patients who felt that an intact leg didn't belong to them mystified him even more than those who complained of pain from an amputated leg. His name for such feelings of alienation added a minus sign to the ghost: 'negative phantoms' – things couldn't get much more spectral.

The term did not catch on. But he had spotted something that Ramachandran made grateful use of in his growing collection of neuro-illusions. Every single one sprang from the interaction between body image and body scheme, between touch sensations and visual impressions, between areas being colonized by neighbouring regions and newly drawn maps, between what the brain has you believe and what mirrors have the brain believe.

Mitchell's ghostly language seems to have been replaced by something that most closely resembles illusionism. But it is precisely here that Ramachandran holds out a respectful hand to his Victorian colleague in the Stump Hospital. If the contours of our body image are indeed so blurry and fluid, shouldn't we perceive the body as a temporary fiction, constantly adapting to changing circumstances? As long as those contours tally exactly with our body, we can cherish the illusion that our body image is etched into our brain. But when they're forced apart by amputations or experimental manipulations, we realize, peering through the resultant gaps, how transitory and elusive they are. To Ramachandran, that was precisely the lesson taught by neuro-illusions: '*Your own body is a phantom, one that your brain has temporarily constructed purely for convenience.*'[48]

FIVE

Whole at Last

In September 1997, a thirty-year-old university lecturer came to see Robert Smith, a surgeon at the Forth Valley Royal Hospital in Scotland.[1] He told Smith that from an early age he had longed to be rid of his lower left leg. He could indicate the precise cut-off point: just above the knee. Though as healthy and sensitive as its counterpart on the right, the lower left leg felt alien, as if it didn't really belong to him. The yearning to be rid of it had got stronger and stronger. Sometimes he found relief in pretence: he would tie it up behind his upper leg and move about on crutches, usually at home, sometimes in public. The relief was only temporary, though: the longing remained. When he was about twenty, he had gone to a psychotherapist, but the treatment hadn't worked. He became severely depressed. The antidepressants he was prescribed did not help: eventually the urge to get rid of his leg became so intense that he tried to burn it, in order to force an amputation. This failed. Desperate, he turned to Smith. After painstaking psychiatric screening, and with the permission of the hospital management, Smith amputated the lower leg at the desired height. Four days later, the lecturer was well enough to be discharged; he had scarcely needed painkillers. He did not experience phantom pain, though every now and again he felt as if his leg were in the same position as when he was pretending it had been amputated.

Two years later, a German businessman in his late fifties came to see Smith with a similar request, though it was the lower right leg that he wanted amputated. In his case, too, the desire had manifested itself at an early age, when he was about thirteen. Initially, the thought of amputation and feigning being an amputee had been sexually arousing but, as he'd grown older, that had diminished. After leaving secondary school, he had gone to work in his father's firm, a company that made artificial limbs. He'd considered having his leg run over by a train, but he had been put off by the fear of other injuries. In his mid-fifties, the desire had become so overwhelming and his frustration about the surplus leg so great that he became clinically depressed. Antidepressants hadn't helped. Smith agreed to operate on him, too. The evening after the operation, he was already walking around on crutches, without painkillers. Five days later, he drove himself back to Germany.

This was the last healthy leg that Smith would amputate. In 2000 he talked about the operations in a BBC Horizon documentary.[2] The TV programme unleashed a storm of publicity, not just from the tabloids. Clare Dyer, legal correspondent to the *British Medical Journal*, wrote an article headed 'Surgeon Amputated Healthy Legs', stating in her opening sentence: 'A surgeon in Scotland amputated the legs of two psychologically disturbed men who had nothing physically wrong with them but felt a "desperate" need to be amputees, it emerged this week.' As a result, the hospital management decided to cancel six similar operations that had been pending.[3]

The two men who Smith had helped represented a larger group suffering from a condition that in 2005 was labelled body integrity identity disorder (BIID), as proposed by the American psychiatrist Michael First.[4] Around nine out of ten people with this disorder are men. They do not have other mental health issues, delusions or hallucinations. Many are clinically depressed, but that is very much the result of unfulfilled longing and the stress of living with an intense desire that usually cannot be shared with others, even close family or partners. Theirs is a lonely suffering. The longing to be rid of the body part – five or six times more likely to be a leg than an arm – usually starts during childhood, is very specific and does not change. A sufferer can almost always remember a specific trigger: one saw a photo of a disabled person as a child; another had encountered someone with an amputated limb and realized with a shock that that was what they wanted too. At home, they experimented with what it would be like to lack an arm or a leg. The longing increases over the years, coming in waves at ever shorter intervals. Finally it becomes so overwhelming that the patient starts to plan ways of ridding themselves of the 'alien' body part. When that succeeds – whether or not with medical assistance – frustration makes way for intense relief. No one has regretted such an operation, even when an entire leg has been removed, or the arm with their dominant hand. A survey of 21 individuals who on average had undergone amputation over eight years previously, often under illegal circumstances, found that every single one was delighted that they had persisted.[5] The difficulty of living with a disability was nothing compared to the satisfaction of having been 'cured'. Their only regret was that they hadn't done this much earlier. Apart from one person in this group who, several years after having his lower left leg amputated, had had a joint of his left index finger removed, none had wanted to lose another body part. For people with BIID, amputation makes the body correspond with how they perceive it to be. They feel cured: paradoxically, the loss makes them feel whole at last.

When choosing a name for the syndrome, First was guided by the analogy with gender identity disorder: just as someone can feel that they are trapped in a body of the wrong sex, a person with BIID can feel that part of their body does not correspond with their body image. Unlike gender identity disorder, though, BIID has not been incorporated in the DSM-5 manual (the standard U.S. classification of mental disorders). As a result, the disorder does not feature in the records of mental-health institutions, and it is difficult for researchers to access data on it. There are no reliable estimates of the frequency with which it occurs. Not all GPs, psychiatrists or psychologists know that this disorder exists and has a name. Many people with BIID only find out in later life – if at all – that there are others like them. The fact that it's so hard to find information on the condition might partly explain why people who have gone to college or university (and are therefore better able to access data) are over-represented among BIID sufferers.

A psychiatrist who has never heard of BIID might conclude that their patient has 'a delusion that their leg does not belong to them' and interpret this as a symptom of psychosis. But not only does the 'delusion' not respond to antipsychotics, patients themselves know all too well that others will find their longings bizarre or perhaps repulsive. Which is exactly why they tend to keep quiet about them. That sense of reality is lacking in the case of a real delusion. A complicating factor, though, is that psychosis can lead to self-amputation. Sometimes a person hears 'voices' in their head instructing them to cut off an arm or leg. The psychiatrist Steven C. Schlozman described thirteen such cases of arm amputation.[6] Matthew 18:8 figures strongly in religious delusions: 'If thy hand or thy foot offend thee, cut them off, and cast them from thee: it is better for thee to enter into life halt or maimed, rather than having two hands or two feet to be cast into everlasting fire.' It goes on, 'And if thine eye offend thee, pluck it out, and cast it from thee: it is better for thee to enter into life with one eye, rather than having two eyes to be cast into hell fire.' The list of body parts that have been cut off, severed, stabbed or sawn off during a psychotic episode is varied and depressingly long: fingers, hands, arms, toes, feet, legs, ears, eyes, tongues, breasts, penises. But self-amputations of this kind result from an overwhelming, uncontrollable impulse – quite unlike the longing, cherished from childhood, to lose a specific body part. If someone were to cut off an arm while psychotic and a surgeon managed to reattach it, the patient would be grateful. A BIID sufferer, though, would be devastated.

BIID support groups and discussion forums are active online. The BIID forum *fighting-it*, set up in 2001, has more than 1,800 members as of September 2024. It supports both people who want to have an amputation (wannabees) and people who like to feign the loss of a limb (pretenders). Then there are the DIYs, people who have developed their own methods to rid themselves of the 'alien' body part. Their accounts, many of them horrific, featuring home-made guillotines, chainsaws, grinders, shredders, firearms, welding guns and injections of poison, are hard to read. Many stop mid-attempt because the pain is unbearable. But that doesn't mean their lives are out of danger. If, after a few hours, a tourniquet is loosened, the toxic substances that have built up in the constricted limb can enter the bloodstream, causing fatal kidney failure. A favourite method is to submerge the limb in dry ice for several hours so that it freezes, thus forcing medical amputation.

In case studies and posts on Internet forums, people with BIID talk about their feelings of loneliness and desperation. No surgeon will help them, but the alternative is so terrible that it isn't really an alternative. When the tension becomes overwhelming, many consider suicide or even attempt it. Even in their letters of farewell they are not open, carrying their feelings of shame and guilt with them to the grave. Some suicides that a person's friends and relatives do not see coming will have been by BIID sufferers who no longer saw a way out. It's also very likely that self-amputations that end fatally are recorded as suicide. In every respect, the statistics of BIID and suicide are a dark number.

A one-legged man's tracks in the snow

There's another, slightly different, mosaic of experiences and perceptions that tends to be placed on the BIID spectrum. Some people have a longing not to lose a specific body part, but to live life as an amputee. Of course this also involves amputation, but the desire isn't prompted by experiencing a body part as 'alien'. There is often an erotic element to the desire to be an amputee. In a candid autobiography, a German chemist wrote about the development of his disorder.[7] Born in 1963, as a child he saw many war veterans and people who'd been disabled by polio. Even back in nursery school, he would watch in fascination, impressed by the dexterity with which they got around on crutches or artificial limbs. He was especially attracted by the sweep of their remaining leg. Later, he started cutting out pictures of disabled people. He was only attracted to men who had lost legs, feeling mainly pity for women or children with the same condition. Once at university, he read more books about orthopaedics than about chemistry, enjoying the illustrations of stumps and prostheses. Every now and again he had girlfriends, but during sex he could only climax by fantasizing about amputees. He would sometimes dream he was paralysed or an amputee, only to wake up, disappointed, to a whole body. After he'd been married for a few years, his wife made him go to a therapist to talk about his sexual problems. The therapist asked what he found arousing. When he told her 'polio victims and one-legged men' she quickly changed the subject.[8] The ignorance and lack of understanding that he encountered among doctors and psychotherapists made him feel even lonelier.

In his mid-thirties, he started to pretend that he had lost a leg. His wife had injured both feet in a horse-riding accident, so there was a pair of crutches in the house. He planned his outings as an amputee as carefully as if he were having an affair. Early in the evening, he would drive to a parking spot next to the motorway, tie up his leg, practise a little with his crutches and then drive to the city. There he would take a tram to the centre and go to a bar, where he would strike up a conversation with someone. People often wanted to know how he had lost his leg and what life was like with only one leg. He remembered these outings as the happiest times of his life. He would even buy shoes, telling the salesgirl that the left shoe was intended for a friend who'd lost his right leg. When people stood up for him on the tram, he didn't feel ashamed, revelling in his temporary identity as an amputee. In winter he would drive to the mountains for a hike, enjoying seeing his own shadow and the tracks of a one-legged man in the snow.

As with many BIID patients, it was the Internet that came to his rescue. In 2006 he read about BIID online and everything fell into place. He realized that he was not alone in his sexual preference, that there were many more people just like him: people whose lives had been burdened by a shameful secret, who, just like him, had sought relief in pretending they were amputees, only to become clinically depressed as a result of their unfulfilled longing. He began to make plans to go abroad to have his lower left leg amputated in an illegal operation. The crucial difference with the other category of BIID patients was that he chose the left leg not because it felt 'alien' but for the practical reason that he would then still be able to drive an automatic. He had already prepared the trip, including a story about how he had supposedly lost his leg, when this option suddenly fell through. He realized he would have to wait until such amputations became legal. His hope was one day to wake up with a bandaged stump and two crutches next to his bed.

As always, it is the erotic element that made everything more complicated. One of the first studies of this disorder, in 1977, concerned two men who each wanted to have a lower leg amputated just above the knee: one the left, the other the right.[9] They had gone to John Money, a psychologist who headed the sexual reassignment programme at Johns Hopkins University. Both had independently reasoned that, just like transgender people, they felt trapped in the wrong body and hoped for an operation to set this right. The psychologist could not help them, but he did record their motives and experiences. One man had made drastic attempts to injure his leg in order to force amputation. In vain, it turned out: staff at the hospital had always managed to save the leg. The other had looked into DIY methods but ultimately shrank from self-amputation. Both men had sexually tinged motives. As one of them put it: 'The image of myself as an amputee has, as an erotic fantasy (each one different), accompanied EVERY sexual experience of my life: auto-, homo-, and heterosexual, since, and beginning with, puberty.'[10] They both collected photos of half-naked amputees to masturbate to and, during sex, fantasized about stumps and prostheses. What aroused them was the image of an asymmetrical body and the thought of caressing the stump.

As far as Money was concerned, the situation was clear. The men were suffering from paraphilia, the label given to sexual behaviours or practices considered abnormal, like fetishism or necrophilia (and, back in the day, homosexuality). Money coined the term 'apotemnophilia' – literally, a desire for amputation. People with this condition, he saw, were aroused by fantasies about sex with an amputee or about themselves as an amputee. The cause, he thought, lay in early childhood experience. For a long time, this psychiatric explanation of the desire for amputation – a sexual deviation, originating in childhood – dominated, as did the term apotemnophilia, only gradually making way for the term BIID.

It had actually been known for a long time that people could become aroused by a stump or paralysis. Back in 1886, in *Psychopathia Sexualis*, his pioneering work on sexual psychopathology, Richard von Krafft-Ebing had written about a thirty-year-old civil servant who could only be sexually aroused by women with a limp.[11] His most intense sexual experiences were with women who were lame in the left leg although, at a pinch, women with a lame right leg would do. In the case of women who were not lame, he was completely impotent. His sexual orientation made him deeply unhappy.

Only the thought of his parents prevented him from committing suicide. He had considered marrying a crippled woman but rejected the idea, feeling it would be dishonest to hide his real motive from her. Von Krafft-Ebing could not help him. Sadly, the medical world couldn't do anything to undo deeply engrained sexual preferences; perhaps it was just better to marry a lame woman.

These days, sex with amputees or the partially paralysed is an established niche on porn sites. On discussion forums for BIID, those who share the German civil servant's sexual kicks are a large and vocal category: 'devotees'. Incidentally, the homepage of *fighting-it* explicitly prohibits the advertising or sharing of devotee materials via the site.

Sexual motives were also cited in larger group studies of BIID. Michael First interviewed 52 people: 5 women and 47 men. Almost 90 per cent indicated that they were sexually attracted to amputees. But when asked if that was the reason for wanting an amputation, only eight men gave this as their primary motive. Forty men said that they wanted to feel 'whole' or 'complete' and that could only be achieved by removing the surplus body part. In another study, nine men with BIID were surveyed.[12] Three stated that they had no sexual feelings for amputees, another three said that sexual attraction was a peripheral element, while the remaining three found the sexual aspect extremely important. So the picture is mixed. The feeling that a limb does not belong to one's body is nearly always the driving motive, but in many cases it overlaps with sexual orientation. Very occasionally, the sexual aspect is of overriding importance. The overlap between motives means that the existence of two kinds of BIID – a 'pure' kind in which a body part feels 'alien' and a sexually motivated variant – is controversial. It has not made the search for an explanation for this disorder any easier.

A reverse phantom limb?

They're mainly legs. They march in a procession of disorders that all have to do with disruption to the sense of ownership. Sometimes it's the direct result of injury. While running away from a bull, the neurologist Oliver Sacks tore the thigh muscle in his left leg. He was operated on, and his leg was encased in plaster. One morning, a nurse woke him up because his leg was hanging out of bed at a strange angle. Half asleep, Sacks assured her that his leg was still lying straight in front of him in bed. She lifted up the blankets. He now saw with his own eyes that his leg was indeed hanging half over the edge, almost at right angles to the position in which he felt it to be. Could she help him put it back in bed? He watched her lift the leg and place it back in the bed but felt nothing. When she'd gone, he refocused his attention on the leg and was suddenly overwhelmed by an intense feeling of alienation. At that moment, Sacks wrote: '*I knew not my leg*. It was utterly strange, not-mine, unfamiliar. I gazed upon it with absolute non-recognition.'[13] He could not make this feeling go away; it only grew stronger. The leg no longer appeared to belong to him; quite the contrary, the leg 'was absolutely not-me and yet, impossibly, it was attached to me and even more impossibly, "continuous" with me.'

This condition (somatoparaphrenia) – luckily temporary in Sacks's case – can become chronic or take on even more extreme forms. Sometimes a person not only feels that the paralysed limb is no longer part of their body, but they actually start to hate it (misoplegia). As a result, they can be driven to mutilate it. If it feels as though the limb belongs to someone else, a patient may make strenuous efforts to shove it out of bed, with the risk of tumbling out after it. In all these disorders, limbs are severed from a person's body image. The sense of ownership is lost. A leg or arm has become 'alien'. But other than in the case of BIID, the feeling is new: it only arises because of an organic disorder and, if that disorder can be cured, the feeling of ownership is restored. The analogy with BIID is faulty, therefore, and studies of these temporary forms of alienation have so far failed to come up with a convincing neurological explanation for feelings of alienation that go back to childhood.

A different line of research has focused not on limbs that were broken or paralysed but on limbs that were perceived as alien by their possessors. After all, those 'alien' arms or legs of BIID sufferers seemed in some ways to be the opposite of phantom limbs. The 'alien' body part is objectively present, but it is not felt to be part of a person's body. Conversely, a phantom limb is objectively missing but, because of the sensations it still seems to cause, remains part of the body image. The former has sometimes been referred to as 'incarnation without animation', the latter as 'animation without incarnation'.[14] In the past few decades, we've learnt a lot more about the brain processes behind phantom phenomena. Might that knowledge help to explain the experiences and perceptions of someone with BIID?

As explained earlier, the thinking now is that phantom perceptions are caused not by stimulation of the nerve endings in the stump (the outdated 'peripheral theory') but by activity in or near the brain area previously connected to the amputated limb (the 'central theory'). Researchers studying BIID were keen to establish whether there were abnormalities in how the 'alien' body part was represented in the brain. Was the lower right leg that the German businessman had Smith amputate connected to the corresponding brain area in the same way as his lower left leg? Was it even connected? Wouldn't it be a neat explanation if it turned out that, in people with BIID, a certain body part wasn't properly 'plugged into' the brain – or plugged in at all – so that it couldn't be recognized as belonging to their body? This has been looked into in studies involving would-be amputees. The idea that 'alien' body parts aren't linked to the brain at all could immediately be rejected. They are controlled in exactly the same way as other limbs. Some differences were found in pain sensitivity, but they were minimal. After the operation, people with BIID reported slightly fewer phantom sensations than patients who had undergone involuntary amputation, but the vast majority of them did have such sensations – further proof that, even beforehand, the brain area corresponding to the amputated limb functioned no differently to that of the non-amputated one.

Some researchers claim to have found abnormalities in the contact between the 'alien' limb and its brain map, but their findings are vague and difficult to assess. After studying four men with BIID, McGeoch, Ramachandran and Brang reported observing abnormalities in the right brain hemisphere tasked with maintaining body image.[15] However, studies of neural substrates (parts of the central nervous system that underlie a specific behaviour, cognitive process or psychological state) typically focus on small groups of test subjects, and the findings prove difficult to reproduce. The abnormalities found are mild; they also vary from study to study. But most importantly, all those graphs and tables measuring the brains of people with BIID leave one crucial question unanswered. To answer it – and even articulate the question – we need first to look at a few more research reports.

A recent brain study – the largest to date – involved sixteen men who wanted to have their healthy left leg amputated.[16] They were recruited via www.bid-dach.org, a German website set up to help people with BIID by providing information to patients, doctors, relatives and therapists. The respondents were compared to sixteen men of the same age and an equivalent level of education. A special questionnaire, the 'Zurich Xenomelia Scale', was used to determine the strength of the men's desire for amputation and the extent to which these wannabes were also pretenders – in other words, how often and how long they had used crutches or wheelchairs to pretend that their leg had already been amputated. The study involved MRI and fMRI scans. The former provide images of brain structure; the latter show how these structures function, registering the brain in operation. All 32 men who took part in the study underwent scans in the radiology departments of the Zürich and Milan university hospitals.

The lower left leg is connected to the somatosensory cortex (the area where bodily sensations are received) in the right side of the brain, the lower right leg to a corresponding spot in the left side of the brain. Initially using the men with BIID as their own control group, the researchers looked at their left and right somatosensory cortices for potential differences. No structural dissimilarities were detected on the right side, though the researchers did find that there was less connection with other parts of the brain (as compared to men without BIID). By contrast, connection was found to be normal in the corresponding areas for the right leg or arms, and in the group of men who did not have BIID. In a more frontal region, responsible for maintaining body image, both structural and functional abnormalities were detected. In this region, the concentration of white matter – the tissue of nerve fibres that acts as the brain's communication network – was less dense on the right than on the left. No differences were found in the control group. There was also a 'gradient': the thinner the concentration of white matter, the longer and more often a patient had simulated lacking a leg.

The researchers interpreted these findings as convincing support for the theory that BIID is caused by abnormalities in the brain circuits responsible for communication with the limbs. Because the longing for amputation arises at such an early age, they theorized, those abnormalities would have been present early on, perhaps from birth – though that would be hard to demonstrate. This would place BIID firmly in the category of neurological, rather than psychiatric, disorders. However, a few caveats are in order.

For a start, these findings haven't yet been replicated, and in this field it's not uncommon for different research teams trying the same experiment to come up with different findings. An earlier study found that sensory stimuli in the unwanted leg were communicated slightly more slowly than those in the desired leg. However, a follow-up study did not detect this difference in processing speed. Because the differences at issue in studies like these tend to be extremely tiny – though they are not described as 'minimal' or 'minor' but rather as 'subtle' – there is lower likelihood of findings being confirmed.

Secondly, the selection of test subjects – in this case men who wanted to get rid of their lower left leg – leads to examination of specific brain areas, where abnormalities are then promptly found. It's debatable whether they would also be detected by researchers tasked with assessing the scans without a preconceived notion of what they were supposed to be looking for.

Much more importantly, though, those differences – big or small, real or imaginary – say nothing about the *direction* of the correlation. If communication between the limb and its neural area is abnormal, could this not also be the *consequence* of many years of abnormal dealings with the 'alien' limb, for example in the form of the patient frequently simulating its absence? The longer and more often a limb is treated differently, the more likely the possibility that this will cause changes in the brain. At the end of their paper, the authors of the study of sixteen men with BIID acknowledge this possibility but at the same time dismiss it as unlikely, concluding that this must be a case of brains influencing behaviour, not the other way round.

Meshugeneh

Most people with BIID themselves believe that they have a disorder. They suffer, they feel abnormal, they regard their longing as an aberration. It comforts them to find out that they are not alone, that there are others with the same disorder. It remains a disorder, though. Many seek help through therapy. In vain: sooner or later they realize that they don't want to be relieved of their longing but of a limb.

But just as an emancipated group of 'Aspies' has formed among people with Asperger's – autists who do not regard themselves as having a disorder but are simply different, 'neurodiverse', and who embrace autism as their identity – some people with BIID regard their desire for amputation not as a disorder but as part of who they are. The fact that they want to have a limb removed, or have succeeded in this aim, is part of their personality, their self – it is what makes them a unique individual. Just like transgender people, they feel trapped in the 'wrong' body but reject the notion of this being seen as pathological.

Michael Gheen (not his real name) is a professor of medicine at a renowned U.S. university who, even as a child, longed to have his lower right leg amputated just above the knee.[17] At most, he views this as 'abnormal' in a statistical sense: it is rare. The difficulties that he and other people with BIID experience are caused not by their desire for amputation but by society's reaction to that wish – with incomprehension and disgust. His position, Gheen believes, is comparable with someone who is Black and living in a racist society, or gay in a homophobic society. The idea that he should seek therapy is completely misplaced: gay people aren't advised to see a therapist: *they* don't have a disorder. That is why homosexuality has been scrapped from the DSM. But although Gheen refuses to accept that BIID is a disorder, he believes it *should* be incorporated in the DSM. Not because it is pathological but because curing this specific form of suffering requires medical assistance. Just as today it is accepted that, after careful screening, transgender people can start a hormonal and surgical trajectory for gender reassignment, so people with a desire for amputation should be helped too. Refusing or prohibiting such assistance in fact amounts to a cruel form of discrimination.

Gheen also revealed that he was one of the four test subjects in the study by Ramachandran, McGeoch and David Brang that allegedly showed a link between BIID and neurophysiological abnormalities. However, he did not see that as grounds for interpreting BIID as a disorder. For one thing, the study does not show whether the abnormalities cause or result from behaviour. And whatever the findings, there's nothing to prevent you from accepting yourself as someone who just happens to have a desire for amputation.

Sometimes it's about more than self-acceptance. The literature on how online discussions help to shape a person's identity as 'someone with BIID' shows that specific amputation desires, once fulfilled, can confer status. One case study quotes patient 'A' who,

by stuffing his support stockings full of dry ice, had forced amputation of both lower legs. As he explained to the doctors treating
him, 'Being an amputee, especially a DAK (double above-knee)
is a rare distinguishing characteristic. It is something about me that
is noticed, accords me certain accommodations and cannot be
denied.'[18] According to him, the double amputation had not only
made him whole but had provided him with an identity that led
to him being respected and accepted.

Even before Gheen and 'A' presented themselves as individuals
with a choice that deviated from the norm but was not therefore
pathological, a debate had already started between ethicists, psychiatrists and doctors as to whether a completely different take
on BIID might be called for. This alternative view was inspired by
the work of Ian Hacking, who documented how, over time, psychiatric diagnoses come into being, catch on, spread and then disappear again. He first demonstrated this transience in a study on
'fugue', or 'mad travel'.[19] People suffering from this condition suddenly become seized by an irresistible urge to roam. They travel
to strange places, often temporarily forgetting who they are, and
afterwards have no memory of the period in which they wan-
dered. The first case – a man from Bordeaux – was diagnosed in
1887, and it was followed by an epidemic of similar cases, spreading from France to Italy and Germany, that would last twenty years.
At the time, psychiatrists believed that the disorder was caused
either by hysteria or epilepsy, the former to be treated with hyp-
nosis, the latter with drugs. Cases of fugue are now extremely
rare, though in 1980 it was still included as a distinct disorder in
DSM III.

As Hacking later also demonstrated in the case of multiple
personality disorder (when someone develops two or more alter

egos), such 'transient mental illnesses' flourish in what he called an 'ecological niche', a spot where the conditions are just right. One such condition is the possibility of linking a disorder to accepted diagnostic categories, as in the case of fugue. Another is the inclusion of a disorder in psychiatric manuals. That only happens when the spread has already started: diagnostic questionnaires or measuring instruments ('scales') are devised to gauge the seriousness of the disorder, and therapies are developed whose costs are reimbursed by health-insurance companies. In the case of more recent disorders, patient associations and discussion forums are set up, and the experiences of people with the same disorder are widely shared online – by this stage the 'epidemic' is a fact.

Back in 2000 the medical ethicist Carl Elliott published an essay in *The Atlantic* with the provocative title 'A New Way to Be Mad' – no question mark – in which he theorized that BIID might also be one of these mental illnesses that appeared relatively suddenly and then disappeared again.[20] He feared that the condition might be 'contagious': the more frequently BIID is presented as a possible identity, one that can moreover confer status within a subculture of wannabes and devotees, the more frequently individuals might be seduced into owning that identity. The amputation of healthy limbs will then no longer be regarded as self-mutilation but as the appropriate medical response to the sufferings of a person with BIID.

A colleague of Elliott's in the field of medical ethics, Arthur Caplan, head of the Division of Medical Ethics at NYU Grossman School of Medicine, regards the desire for amputation as hardly worth discussing: 'It's meshugeneh – absolutely nuts. It's absolute, utter lunacy to go along with a request to maim somebody.'[21] To him, that would violate the Hippocratic Oath and the principle of doing no harm. But more importantly, the mere fact of someone wanting such an operation proved to Caplan that a wannabe is not competent to be involved in decisions of this kind.

Neurologists such as Leonie Hilti and Peter Brugger reject the idea of BIID as a freely chosen identity. BIID is no 'new way to be mad', they wrote in response to Elliott; it is a disorder associated with demonstrable cerebral abnormalities.[22] A set element in disputes about whether or not a disorder is transient is the argument that the disorder has always existed, but it has only now been identified and is therefore unjustly being labelled as a fashionable disease. In the 1990s, for example, the boom in repetitive strain injury (RSI) diagnoses led to claims that medieval scribes also suffered from RSI. A historical argument is also mooted in the discussion about BIID. Curiously – but wait for it, things are about to get much weirder – it was put forward by the same Carl Elliott who regards BIID as a transient mental illness.

In an article in a medical ethics journal that he wrote together with the lawyer Josephine Johnston, Elliott refers to a story recorded in 1789.[23] It featured in a collection of tall tales by doctors and surgeons, collected by an anonymous anatomist, possibly the physician Jean-Joseph Sue or his brother Pierre.[24] In 1781 or 1782 an Englishman was said to have offered to pay a French surgeon the princely sum of 100 guineas to amputate his leg. At first the surgeon refused. The leg was perfectly healthy, in excellent shape – it would be wrong to cut it off. But the Englishman drew a pistol on him, stating that he would not hesitate to use it if he did not get his way. The surgeon protested that he didn't have the proper equipment to perform such an operation. The Englishman appeared to have anticipated this: besides the pistol, he had brought with him surgical instruments and bandages. With the gun still pointed at him, the surgeon capitulated. He cut off the leg. The man made a good recovery and travelled back to London with a wooden leg. Sometime later, the surgeon received a letter from this curious gentleman ('*cet original*') containing a money order for 250 guineas. 'You have made me the happiest of all men,' explained the Englishman, 'by taking away from me a limb which put an invincible obstacle to my happiness.'[25]

So far, the story ties in with what, two centuries later, would become the standard version of BIID: an intense longing for amputation, followed by deep satisfaction when this goal was achieved. And this is the point up to which Johnston and Elliott cite the account. But if you look up the original story, you see that it carries on and that the case takes a very different turn. In the rest of his letter, the Englishman explains the reason for his strange request. He wrote that he was madly in love with a one-legged woman. He had asked her to be his wife, but she had turned him down. She said she didn't want to feel inferior to her husband because of this disability, fearing that he might later reproach himself for having married her. Although he swore that he loved her for who she was, she continued to refuse. So he had taken the drastic step of (literally) placing himself on an equal footing through amputation. Back in London, he had once again proposed to her, and she had accepted. They were now happily married. The Englishman was no eighteenth-century case of BIID; he didn't want to be rid of an 'alien' leg; he had crossed the Channel to make a sacrifice on the altar of love.

Why did the authors only quote the first half of this tale? When questioned, Elliott said that a psychiatrist had given him a translated version – he himself couldn't read French. The story, misleadingly shortened, went on to lead a life of its own. Hilti and

Brugger invoked the case of the Englishman as proof that BIID was a disorder that went far back in time.[26] Chopped in half and taken out of context, the story continues to appear in numerous articles, cited, as it is put, from 'a medical text'. A tale (already questionable) from a collection of medical yarns has thus been turned into an argument for the cerebral origin of BIID.

At present, convincing 'early' cases of this disorder – like the centuries-old ones in the case of Cotard's or Capgras – appear to be lacking. They may turn up, of course, but for the time being, arguments against the notion that BIID is spreading because of the 'contagiousness' of sufferers, either online or elsewhere, must be sought elsewhere. People with BIID have often only found out via the Internet that there are others with the same intense longing to be rid of an 'alien' limb. That longing wasn't something they were talked into – for that they were too young, the disorder was too rare and information too scarce. The accounts they independently give show great consistency: the early start, the trigger at a young age, the increase in intensity as the person ages, the futility of therapy, the onset of depression when the outlook looks hopeless, the temporary relief through feigning disability, and the joy and absence of regret if they manage to obtain an amputation. The fact that most keep their desires secret for so many years is one of the reasons why it has taken so long for 'standard accounts' – personal histories in which others might feel affinity – to circulate.

The emergence of Internet discussion forums, newsletters and chat groups has changed all this. These days, hundreds of life stories of people with BIID have been posted online. Anyone who wanted to could easily assemble elements from these prototypes and appropriate BIID as an identity. Whether this is happening, and if so on what scale, is unknown and hard to research. If it *is* happening, it would be this group that Elliott sees as falling victim to 'contagion' and paying the price in the form of a limb. But crucially, no such prototypes could have existed in the first place had there not been a well-documented community of people who – from an early age and without being aware of others in the same position – had a burning desire to make their body correspond with their body image.

SIX

Grief Hallucinations

When the numbness that gripped you in the first days and weeks after your loss starts to wear off, it's as if your memory has meanwhile gone on strike. It downs tools, leaving you floundering. All of a sudden you can't recall a password that you typed in blindly for years. In the supermarket, you forget what you went out to buy. You can see from the pitying look of a person you're talking to that you're repeating yourself. Sometimes it's an effort to remember what day of the week it is, or who came to see you yesterday. The simplest tasks take twice as long as usual. Absentmindedness, inability to concentrate, forgetfulness: these are the standard ingredients of grief. It's a time when your memory stumbles around, as the Dutch writer W. F. Hermans put it, 'like a drunk manservant who, sent to fetch a bottle of vintage wine from the cellar, comes back only with spiderwebs, shards of glass and tales of ghosts.'[1]

The memories that do surface, whether bidden or unbidden, conjure up feelings that can be so conflicting you really don't know what to make of them. You cherish memories of your lost loved one, and it's comforting to be able to swap recollections with close friends or relatives. But you start to realize that the loveliest memories – the ones you hoped would later help you bear your loss – are now the most painful. They illuminate what has gone, shine a spotlight on an absence, mark what cannot be recovered. They've become memories that you alternately try to recall and shrink from.

You also begin to discover the power – and peril – of association. The house that you shared is one great network of associations, every single one a reminder of the person you loved. Rooms, furniture, belongings, household tasks, sounds, smells, the way the light falls at certain times of the day: literally everything in the house highlights what is no longer there. That's why some people literally flee their homes – into the streets, a park, to friends – to escape the relentless onslaught of memories. Others decide to move.

While maddeningly fuzzy in many ways, your memory can summon up old arguments and conflicts with laser-sharp clarity. You feel guilty and ashamed about your own part in them, but you can't undo anything. Conversely, you can no longer forgive the other person for doing things that annoyed you, or for which you reproached them. You can't speak about grievances, or even dwell on them silently yourself, without feeling disloyal. Even the first lessening of grief, months and months after a loss, can prompt feelings of guilt and thus actually prolong mourning.

Grief is a time of outward silence and inner tumult. And this chaos can generate two conflicting experiences, both relating to memory.

The first, shortly after bereavement, is a disconcerting gap in your memory: you suddenly realize you can no longer recall the face of your loved one. You've woken up, it's dark, you're trying to conjure up that oh-so-familiar face – and nothing appears. You feel a surge of panic. If I already can't remember their face, what else will I have forgotten in a few weeks or months? During the acute stages of grief, experiences like these are not uncommon. Luckily, they're a fleeting phenomenon – but that doesn't make it any less frightening.

By contrast, the second experience is almost universally perceived as comforting. Every now and again, your memory lets you once again see or hear your loved one in the form of a hallucination. You suddenly see your husband sitting in his usual chair, you hear your wife say something or her footsteps coming down the stairs. More than half of people who have lost a loved one have experiences like these. In the older literature, they used to be known as 'hallucinations of widowhood'. No one was particularly happy with the term. Not only do they occur just as often to widowers, the apparitions can also include children, close relatives or friends who have died. So 'grief hallucinations' is a much better term.

Grief hallucinations are common, but at the same time they are a well-kept secret. People who have these experiences tend to keep them to themselves. GPs usually only hear about them when a concerned relative insists that the person see a doctor. So, what do we know about grief hallucinations?

A sense of presence

In 1817 the French psychiatrist Jean-Étienne Esquirol defined hallucinations as the sensory perception of something that is not there. He distinguished them from illusions, which involve wrongly interpreting a sensory stimulus – like when you see a shadow and think it's a dog. In the case of hallucinations, there is no stimulus. Emil Kraepelin, the 'Linnaeus of psychiatry', whose classification system for types of psychiatric disorder is still largely in use today, classed hallucinations as a symptom of psychopathology. But associating hallucinations with pathology is unfortunate, because they can occur independently from any kind of psychiatric disorder. Grief hallucinations are a good example: there is absolutely no link with psychiatric or neurological disorders. The literature has proposed more neutral-sounding alternatives, like post-bereavement experiences (PBES). They are also seen as a possible indicator of persistent complex bereavement disorder, a condition that – according to the DSM-5 – people are deemed to be suffering from if they don't 'complete' the grieving process within a set space of time.[2] The 'natural' grieving process should be all wrapped up within six months, just so you know. But let's not go down that road.

The pioneering work on what was then called 'hallucinations of widowhood' was carried out in 1971 by William Dewi Rees, a GP working in a rural practice in mid-Wales.[3] He interviewed 293 people who had lost their partners – 227 widows and 66 widowers. Both groups were eager to discuss their experiences.

Almost half of those questioned had had hallucinations, men approximately as often as women. The most common experience – shared by 40 per cent – was the feeling that the dead person was still there. This was followed by seeing (14 per cent) or hearing (13 per cent) the dead person. Seven women and one man had had the feeling that the dead partner had touched them, but this sensory hallucination was rare: under 3 per cent. Around one in ten of those questioned still spoke to their partner, men more frequently than women. Widows tended to hear their husband more frequently – for example their voice or their footsteps – whereas widowers more often reported seeing their wife.

Relatively young widows and widowers (under sixty years old) had far fewer hallucinations or conversations with the dead partner. In the first ten years after a loss, hallucinations and conversations were more frequent than ten or twenty years later. The longer a marriage, the greater the likelihood of hallucinations.

If the marriage had been childless, widows and widowers were less likely to experience hallucinations. Eleven widows indicated that their marriage had not been happy: they did not have any hallucinations. (Contrary to what this might suggest, absence of such hallucinations did not indicate an unhappy marriage.) Hallucinations appeared to discourage remarriage: four of those questioned had turned down a proposal because they felt that their dead partner was against it.

There were also factors that did *not* have any impact on whether or not people had hallucinations. Whether the death had been sudden or expected; whether the person had died at home, in hospital or elsewhere; whether someone lived in a village or city; was religious, agnostic or an atheist; was lonely or supported by family and friends; remained in the same house after the death or moved home – none of this made any difference.

Up until then, most of those interviewed had not discussed their hallucinations with anyone. They were inclined to keep quiet about them, widowers more so than widows. No one had spoken to their GP about them. When asked why they had been so reticent, around half could not give a clear reason. The rest had varying explanations: they were scared others would think it ridiculous; it was too personal; no one had asked them about it; it wouldn't interest anyone else; people would think they'd lost their marbles. As a result, these remained private, intimate experiences that the vast majority perceived as comforting and supportive at a time of sorrow. Only one in twenty of the bereaved would rather not have had the hallucinations.

Subsequent research largely confirmed Rees's findings. P. R. Olson and colleagues interviewed a group of about fifty widows (with an average age of eighty) in two retirement homes in Asheville, North Carolina.[4] When the women were asked if they sometimes had the feeling that their husband was still with them, more than 60 per cent said that they did, usually in the form of a visual hallucination (almost 80 per cent). Hearing their husband's voice or some other sound made by him was somewhat less common (50 per cent). The feeling of being touched was rare. Unlike the cases reported by Rees, childless widows had slightly *more* hallucinations. Many of the women had been widowed for quite some time, and the positive correlation that Rees had found between the length of a marriage and the likelihood of hallucinations turned out to be somewhat more complex in North Carolina. Most hallucinations occurred in two age groups: women who'd been widowed between the ages of thirty and forty, and women who'd lost their husband after turning seventy.

One of the questions was whether they still dreamt about their husband. Most did (85 per cent), largely the same widows who had experienced hallucinations. Of the seven widows who never dreamt about their husbands, only one had had hallucinations. Just as reported by Rees, nearly all widows viewed the appearance of their husband in hallucinations as a positive, reassuring experience – that they had nevertheless kept to themselves.

In 2006 more than eight hundred widows and widowers (aged between 65 and 80) in the Danish city of Aarhus were asked to take part in a survey.[5] The final questionnaire in the survey was sent over four years after they'd been bereaved and, just as in

Rees's case, met with an unusually high response rate: three out of four widows and widowers replied. On average, they had been married for 44 years. More than half those questioned said they had experienced hallucinations, which most of them perceived as comforting. Only 7 per cent regarded it as a negative experience. This study, too, found that the longer a marriage, the greater the likelihood of hallucinations.

It's as if my insides had been torn out

At first you feel numb, it doesn't really hit you – In the beginning you go through life in a daze – A feeling of disbelief: surely this can't have happened – Gradually you begin to grasp how much your life has changed – The extent of the loss only becomes clear when you realize just how much is now impossible – I no longer felt whole – It's as if my insides had been torn out – You often think back to how things were before the loss – A feeling of anger, why did this happen to me, how come other people can just carry on with their lives?

These are fragments of conversations, not with widows or widowers, but with people who have had a limb amputated.[6] Colin Murray Parkes, at the time a consulting psychiatrist at St Christopher's Hospice in London, noticed that in the first few months after the loss of an arm or a leg, patients tended to express themselves in the same terms as people who had lost a loved one. People who first lost a limb and later their partner – or vice versa – indicated that their feelings and the order in which they occurred were very similar.

As if to offer a literary illustration of Parkes's observations, Paul Auster, in his novel *Baumgartner*, has his protagonist Sy Baumgartner, a seventy-year-old retired professor whose wife Anna drowned after being hit by a rogue wave, muse along the same lines. He even studies the literature on phantom experiences – Silas Weir Mitchell, Oliver Sacks, Vilayanur Ramachandran – only to conclude that amputation is possibly the most powerful metaphor of his grief:

> It is the trope that Baumgartner has been searching for ever since Anna's sudden, unexpected death ten years ago, the most persuasive and compelling analogue to describe what has happened to him that hot, windy afternoon in August 2008 when the gods saw fit to steal his wife from him in the full vigour of her still youthful self, and just like that, his limbs were ripped off his body, all four of them, arms and legs together at the same time, and if his head and heart were spared from the onslaught, it was only because the perverse, snickering gods had granted him the dubious right to go on living without her. He is a human stump now, a half man who has lost the half of himself that had made him whole, and yes, the missing limbs are still there, and they still hurt, hurt so much that he sometimes feels his body is about to catch fire and consume him on the spot.[7]

This echoes what Parkes heard when he interviewed amputees and widows and widowers.

A 51-year-old woman who had been widowed fifteen years previously and later lost her right arm in an accident at work said: 'It was when I came home, when people stopped coming, when I got used to being here alone, that realization hit me. It was the same with my husband dying. I felt that feeling of loss. As though it was the end of the world. You don't know what to do. Forlorn – hopelessness – it was like grief, more or less. Why did God let it happen? I had the same feeling when Bert died.'[8] Numbness, haze, disbelief, constantly being reminded of your loss, struggling to comprehend its irreversible nature, an ever-growing list of activities and pursuits through which a line has now been drawn, a future that constantly needs rethinking, bitterness – they are stages in two bereavement processes that mirror each other.

The feeling that amputees share with widows and widowers is one of continuing presence. In the case of the former, this takes the form of phantom sensations, in the case of the latter, grief hallucinations. The person – or thing – that has vanished has not really gone because they still appear as images, sounds, voices or sensations. People who have experienced both amputation and the death of a loved one feel the echo of their first loss when the second occurs. Grief hallucinations remind them of the period just after amputation. Conversely, phantom sensations have the same effect on those who have previously been bereaved.

But the emotional impact of grief hallucinations is quite different from that of phantom sensations. Grief hallucinations are comforting; they seem to keep the dead person with you for a little while, to postpone the farewell, make the loss less final. They transport you, however briefly, back to life before your loss. Phantom sensations, by contrast, are usually unpleasant – like an itch, pain or cramp – and even more neutral sensations, like the feeling that your phantom limb is bent or stretched, are not perceived as comforting. Sensations from a missing limb do not soften your loss, they confront you with it. A second difference is that grief hallucinations are rarer than phantom sensations and don't last as long. Thirteen months after amputation, more than half of patients still occasionally had a 'sense of presence'; in the case of widows/widowers, this applied to only one in seven.[9]

An emotion that's unique to amputees is the concern many of them have about what happened to their amputated limb. For someone who loses a loved one, there are rites for what happens to the body after death (washing, dressing and laying out), as well as rituals surrounding funerals and cremations. But there is little understanding for anyone fretting about the fate of their amputated leg. Around half of patients wondered what had happened to their limb after amputation, and a third had been really bothered by the question. A woman was quoted as saying: 'My husband said, "Don't be silly, they burn it." I thought, "There's a part of me being cremated." It seems silly, doesn't it?'[10]

A painful parallel in life both after amputation and after bereavement is the effect that reflexes can have. After an amputation, a limb can seem so present that the person simply forgets it's not there. They get up out of their chair to fetch something, realize their mistake too late and fall over. Or they might reach out to catch a falling cup, only to realize that their arm has gone. They are reflexes that confront patients painfully with their loss. In the first weeks and months after the death of a loved one, reflexive behaviour can spark the same confrontations with the new reality. Again, Auster presents an illustration. It is only ten days after the funeral when Baumgartner reads something that he is sure will interest Anna. 'In a blur of forgetfulness', writes Auster, he walks toward where he expects to find her – 'and then he stepped into the empty living room and remembered'.[11] Sadly, lapses of memory like these are common experiences. Out of force of habit, you lay an extra place at the table for your partner. Something happens that makes you think 'I must tell this story when I get home.' Or the phone rings, and you expect the other person to answer it. It only takes a few seconds for you to correct yourself, but such moments are a constant, cruel reminder that nothing will ever be the same again. It's as if the impact of your loss doesn't hit you all at once but is parcelled out over time.

Windows onto conflicts

Grief hallucinations aren't only experienced by people who have lost their partners. A biographer of Silas Weir Mitchell described how Mitchell had heard from a journalist that one of his best friends, Phillips Brooks, had died unexpectedly.

> Mitchell, greatly shaken, went up to tell his wife. On the way back downstairs he had an odd experience: he could see the face of Brooks, larger than life, smiling, and very distinct, yet looking as if it were made of dewy gossamer. When he looked down, the vision disappeared, but for ten days he could see it a little above his head to the left. The strange incident carried his mind back to the mysterious footsteps heard by his father and mother on the night of his brother Alexander's death.[12]

A woman who was undergoing therapy for panic attacks related how her father had died unexpectedly of a heart attack nine months earlier. In the six weeks after his death, she would sometimes walk in the park where her father had liked to stroll. On three occasions she had seen him clearly, sitting on a bench, as he used to do. She halted, looked at him, turned and walked a little way off, then looked back once more and saw that he had disappeared.[13] Unlike most grief hallucinations, this woman perceived her father's apparition as sad, and in future she avoided that path.

Feelings of guilt and regret can resonate in grief hallucinations. Unintentionally, they become windows onto conflicts that weren't discussed or were left unresolved while the other person was still alive. Sixty-nine-year-old Mrs P. lost her husband to a heart attack.[14] His wish had been to die at home, in his favourite rocking chair, but his wife wanted him to stay in bed. When he got up and tried to make his way to the rocking chair, he had fallen. He had to be taken to hospital, where he died twelve hours later, though not from the consequences of the fall. Shortly after the funeral, Mrs P. started to have hallucinations. She saw him sitting in his rocking chair, occasionally smelt his aftershave or heard him pottering about the house. She told the district nurse and her GP that the apparitions seemed restless, and she could imagine why: after all, it was partly her fault that he had not died at home. Both assured her time and again that she had taken sensible decisions during his illness and cared well for him. The apparitions continued though they became less frequent, gradually shedding the restlessness that had so upset her.

Some grief hallucinations are so intense that they can be dangerous. A woman in her forties presented herself at a psychiatric polyclinic in Berlin.[15] Seven months earlier, she had lost her drug-addicted daughter to an overdose. A few days after her death, she began to have hallucinations of her daughter, initially in the form of images and smells, later as sounds. The woman heard her calling 'Mama! Mama!' or 'It's so cold.' This happened two or three times a week, and the hallucinations could last for a few minutes. They were so real that the mother occasionally replied. She had previously been prescribed the kind of antipsychotics used to treat schizophrenia, but they had not helped at all. The hallucinations occurred at unpredictable moments, and it was this that worried her: she feared that her daughter might suddenly appear when she was driving, and that the shock might cause an accident. A combination of sedatives, antidepressants and therapy to deal with her feelings of guilt did not end the hallucinations, but the fear and horror lessened.

Sometimes hallucinations appear to worsen a relationship that was troubled during life. A young woman ('Julie') had an extremely difficult relationship with her mother.[16] Her older brother had been the favourite at home, and she felt herself to be unwanted. She had found out that she had been named after a woman with whom her father was having an affair. Julie had then changed her name, but her mother refused to call her by it. After her mother's death, she started having nasty hallucinations: she would hear her mother abusing her ('slag', 'slut' and 'whore') or telling her that she wasn't fit to live. These were not flashbacks or memories because, during her lifetime, her mother had never said things like that to her. She had the impression that the hallucinations were giving voice to an enmity that had remained unspoken while her mother was alive.[17] The relentlessness of the hallucinations caused her to start to doubt herself ('When you hear a thing often enough . . .'). What's more, the sense of rejection that she had always felt was now heightened.

After the death of her boyfriend, another woman ('Aggie') was left nursing a grievance that she could do nothing about. The boyfriend had had a serious heart condition that he had concealed from her. When he found out he was terminally ill, he had ended their relationship in an attempt to spare her pain. Aggie couldn't understand it at all; she felt spurned and went through a miserable six months. Shortly before his death, there was a partial reconciliation: he could not do without her after all. In the hallucinations he would say 'I'm sorry', something he had never said to her while he was alive. That was what bothered her so much, that he had never apologized for ending their relationship. His posthumous apologies made it somewhat easier for her to deal with her own feelings of grievance: 'Now I understand why things happened.'[18]

The accounts of grief hallucinations experienced by Julie and Aggie come from a study in which most hallucinations – in twelve of the fourteen cases – were positive. In other studies, too, positive hallucinations dominate. A widow told how, one night in bed, she felt a tap on her shoulder and heard her husband say: 'It's all right love, it's only me, but everything is fine.'[19] She had felt this to be reassuring: 'I thought, well perhaps he's telling me it's okay, you know, get on with life and don't worry.'[20] There are many examples of grief hallucinations that have a similarly reassuring effect. Pat, a forty-year-old woman who had lost her beloved aunt, Enid, said that she had once been so overcome by the loss that she had burst out crying in the middle of a crowded bookshop, much to her embarrassment. 'I felt a hand on my back, which was what Enid always used to do, when I'd leave she'd just put her hand on my back – kind of gently stroke – she wasn't a huggy sort of person and I just felt this stroke and I just heard "it will be ok" – and things just kind of came to – all the rage and the rawness and everything else just like a deflated balloon. I don't know if I've conjured her up – it doesn't really matter – it works.'[21] That question about conjuring up is something that recurs in accounts of grief hallucinations. Seventeen-year-old Sandra lost her boyfriend. In the first stage of bereavement, she continued to feel his presence.

'He hadn't left yet, so I had him for a little bit longer. I kind of felt I know he was gonna go, that his soul wasn't gonna be around forever, but I kind of felt he hadn't just gone, just totally disappeared . . . And sometimes it sounds really loony and I'll think am I making this up – did my mind just construct something because it helped?'[22] The doubts expressed by Pat and Sandra about their own share in the occurrence of hallucinations raise tricky philosophical questions. Can you imagine something that surprises you, as in Pat's case? Is it beyond your control? And if so, is it produced by your mind, as Sandra suggests? Or by your brain?

There is a category of hallucination that differs from grief hallucinations, yet at the same time is sufficiently similar to be regularly confused with them, even by doctors. The explanation of these 'Bonnet images' could also shed light on the origin of grief hallucinations.

Emergency rations

In 1760 the Swiss naturalist Charles Bonnet was the first to identify a curious category of hallucination.[23] After undergoing cataract operations on both eyes, his grandfather, nearly ninety, began to see strange things. Once, when his granddaughters were visiting him, two young men walked into his room. They were wearing magnificent cloaks, and their hats were trimmed with silver. His granddaughters swore that there was nothing there. Another time, some elegant ladies passed by, some with a small box on their head. The images were extremely detailed, and they all glided past silently, even when he could see that the people were talking. By covering his left and right eye in turn, he established that they were not caused by wrongly interpreted visual stimuli – even when he covered both eyes, these images still occurred.

Bonnet did not refer to the images as hallucinations but as 'visions' – that is, something that was viewed. They were not 'symptoms' as far as he was concerned; his grandfather was not 'suffering' from these images, nor was he a 'patient'.[24] The medicalization of experiences like these only happened much later. In 1936 Georges de Morsier, a neurologist from Geneva, suggested that the visions be classified as hallucinations and anyone experiencing them as suffering from 'Bonnet's syndrome'.[25] Today, the images are regarded as completely harmless, even though they are usually caused by problems with sight. Robert Teunisse, a specialist in geriatric psychiatry, has clarified much about the phenomenon.[26] The images are extremely varied, sometimes appearing in colour, sometimes in black and white, sometimes sharp, sometimes blurry, sometimes lasting more than an hour, sometimes disappearing after a few seconds, sometimes of normal size, sometimes miniaturized or extremely elongated. The images mostly appear at dusk, in the evening or night, often at a time of inactivity. Tiredness also makes them more likely to appear. The statistics in Teunisse's study can be used to generate a kind of composite portrait:

> The prototypical Bonnet patient who emerges is someone well on in years, with severely impaired vision. Reading is almost impossible. His surroundings tend to be quiet. He lives alone, and lacks both the energy and the inclination to leave the house or to look up old friends. He receives few visitors. His days pass by calmly, each one as uneventful as the previous one. Towards dusk, when he starts to feel a little drowsy and the outside world is beginning to fade, the images appear. He does not find them disturbing, since he knows they aren't real. And he can make them

disappear: all he has to do is blink his eyes and they're gone. But he doesn't talk about his experiences. After all, they are a bit strange. He doesn't want people to think he's losing his faculties.[27]

Fear of being thought mentally unstable is shared by people who have grief hallucinations. But there are other similarities. Grief hallucinations and Bonnet images appear spontaneously; they cannot be summoned up. Nor can they be prolonged at will. Their content can't be controlled. People who experience them immediately realize they aren't real, so they are seldom perceived as threatening. Unlike memories or daydreams that take place 'in your head', Bonnet images and grief hallucinations are projected on the outside world: they feel like a sensory experience. Sometimes doctors prescribe antipsychotics in both cases, but it does not make the images disappear. Nor do they respond to other drugs, like antidepressants. And even if effective drugs did exist, almost no one would want to take them: they don't want to be rid of the hallucinations. Neither of these types of experience is indicative of a psychiatric or neurological disorder. Patients merely need to be told about their innocent nature to be sufficiently reassured.

Of course, there are differences too. Bonnet images are put together from the imagination – they contain scenes and figures that were never seen in real life – whereas grief hallucinations, by contrast, centre on familiar people, events or conversations. Bonnet images are almost always associated with a period of impaired vision and usually clear up soon afterwards; grief hallucinations are not linked to problems with sight, involve other senses as well and persist for longer periods. The perception of Bonnet images is neutral, that of grief hallucinations emotionally charged. But the analogy between the two suggests that they might share an explanation.

As far as we know, Bonnet images are a reaction by the brain to a lack of visual stimuli. Whatever the cause – cataracts, glaucoma, infections, a detached retina – the visual areas of the brain receive fewer and fewer stimuli to process. And the brain does not respond well to such unemployment. By way of compensation, it starts to manufacture its own images and offer them to view. The fact that they seem to come from an external source is logical: that's how people are used to perceiving images. In reality, the images are a kind of emergency ration with which the brain keeps itself active. And it only feels the need to do this during the first stage of visual impairment: by the time someone's vision has deteriorated further, the brain has adapted to the new situation. Once someone is completely blind, the Bonnet images stop. The whole cycle of appearance, diminishment and disappearance of such images is entirely outside a person's control. It's something your brain has you experience.

Might there be comparable neurological mechanisms that can soften an intense emotional loss? An intervention by your brain in response to a shocking bereavement? All those associations that cause you to continually bang up against the brick wall of loss – might not your brain sometimes compassionately dress them up into something you think you see or hear in the external world? For this to work, such hallucinations mustn't feel like the product of your imagination: they must appear to be external. And they must be of a certain type. It's not for nothing that grief hallucinations that worsen and prolong grief are an exception. The vast majority provide consolation and moral support; they encourage, reassure, tell you you're on the right track, help you when you're in doubt, soften feelings of remorse and regret. The fact that grief hallucinations give you what you hope (and only very occasionally fear) is perhaps the clearest pointer to their origin.

But can you really be comforted by yourself? It's an ambiguity that was nicely described by the writer Elias Canetti. Canetti was good friends with the composer Alban Berg, who died at the age of fifty. In his autobiography, Canetti wrote that at a lecture some thirty years after Berg's death, he met his widow Helene. She was then in her eighties, a small, frail woman. Canetti hesitated to speak to her but did it anyway. At first she didn't recognize him but, when he introduced himself, she said: 'Oh, Mr C! It's been a while. Alban still talks about you.'[28] Canetti was so moved by this that he quickly took his leave of her and afterwards could not bring himself to visit her at home.

> I didn't want to disturb the intimate conversation in which she remained caught up, it was as if their entire past life together was simply continuing to this day. In matters relating to his work she would ask him for advice, and he would give her the answers that she herself had come up with. Does anyone believe that others knew his wishes better? It takes a great deal of love to recall a dead man to life so fully that he never vanishes, that you hear him speak, talk to him, learn his wishes, which he will always have, because you yourself have created him.[29]

REFERENCES

Through the Looking-Glass

1 L. Murat, *L'Homme qui se prenait pour Napoléon* (Paris, 2011), p. 53.
2 P. Pinel, *Traité médico-philosophique sur l'aliénation mentale, ou la manie* (Paris, 1800/1801), quoted in Pinel, *A Treatise on Insanity* (London, 1806), p. 70.
3 E. Nourry, 'Les Saints céphalophores. Étude de folklore hagiographique', *Revue de l'histoire des religions*, XCIX (1929), pp. 158–231.
4 S. Laskow, 'The Decapitated Saints Who Still Managed to Hold Their Heads Up', *Atlas Obscura*, www.atlasobscura.com, 30 October 2015.
5 Pinel, *Treatise*, p. 72.
6 Pinel quoted in Murat, *L'Homme*, p. 56.
7 Ibid., p. 104.
8 Ibid., p. 105.
9 Ibid., p. 163.
10 L. Carroll, *Through the Looking-Glass (And What Alice Found There)* (London, 1871).

ONE The Three Christs of Ypsilanti

1 The search term 'Ann Arbor News Ypsilanti State Hospital' brings up more than a dozen photos taken in Ypsilanti in 1937.
2 J. D. Thiesen, 'Civilian Public Service: Two Case Studies', *Mennonite Life*, XLV/2 (1990), pp. 4–12: p. 8.
3 C. Zbrozek, 'Opening Its Doors Again', *Michigan Daily*, 28 September 2005.
4 R. Christie, 'Milton Rokeach, 1918–1988', *American Psychologist*, XLV/4 (1990), pp. 547–8: p. 547.
5 L. Festinger, *A Theory of Cognitive Dissonance* (Evanston, IL, 1957).
6 M. Rokeach, *The Three Christs of Ypsilanti: A Narrative Study of Three Lost Men* (New York, 1964). All unacknowledged quotations in this chapter are from Rokeach's book.

7 R. Lindner, *The Fifty-Minute Hour: A Collection of True Psychoanalytic Tales* (New York, 1958), pp. 193–4.

8 D. Capps, 'Identity with Jesus Christ: The Case of Leon Gabor', *Journal of Religion and Health*, XLIX/4 (2010), pp. 560–80.

9 H. H. Perlman, 'Book Review: The Three Christs of Ypsilanti', *Social Service Review*, XXXVIII/2 (1964), pp. 236–7: p. 237.

10 B. Nachmann, 'Book Review: The Three Christs of Ypsilanti', *International Journal of Social Psychiatry*, XI/4 (1965), p. 313.

11 M. Rokeach and N. Vidmar, 'Testimony Concerning Possible Jury Bias in a Black Panther Murder Trial', *Journal of Applied Social Psychology*, III/1 (1973), pp. 19–29.

12 R. Moody, 'Introduction', in M. Rokeach, *The Three Christs of Ypsilanti* (New York, 2011), p. xiv.

13 J. Diski, 'Diary: The Three Christs of Ypsilanti', *London Review of Books*, XXXIII/18 (22 September 2011).

14 'The Three Christs of Ypsilanti', *New York Public Radio*, www.wnyc.org, 24 March 2017.

15 P. Pinel, *Traité médico-philosophique sur l'aliénation mentale, ou la manie* (Paris, 1800/1801), quoted in Pinel, *A Treatise on Insanity* (London, 1806), p. 97.

TWO Living in the Knowledge of Being Dead

1 H. Förstl and S. Dickson, 'Paramnestic Reduplication of Charles Bonnet and Theophile Bonet', *British Journal of Psychiatry*, CLX/3 (2021), pp. 416–18.

2 G. Mak, *The Many Lives of Jan Six: A Portrait of an Amsterdam Dynasty*, trans. Liz Waters (Amsterdam, 2017), p. 131.

3 W. J. Maas et al., eds, *Memori Boeck. De wereld van Cornelis Pieter Sorgdrager (1779–1826)* (Hollum, 1983), p. 1.

4 J. Cotard, 'Du Délire hypochondriaque dans une forme grave de la mélancolie anxieuse', *Annales Médico-Psychologiques*, IV (1880), pp. 168–74. An English translation appears in G. E. Berrios and R. Luque, 'Cotard's "On Hypochondriacal Delusions in a Severe Form of Anxious Melancholia"', *History of Psychiatry*, X/38 (1999), pp. 269–78.

5 Cotard, 'Délire hypochondriaque', p. 168.

6 Ibid., p. 169.

7 J. Pearn and C. Gardner-Thorpe, 'Jules Cotard (1840–1889): His Life and the Unique Syndrome Which Bears His Name', *Neurology*, LVIII (2002), pp. 1400–1403.

8 J. Pearn, 'A Biographical Note on Marcel Proust's Professor Cottard',
   Journal of Medical Biography, XI/2 (2003), pp. 103–6.

9 A. Ritti, 'Éloge du Docteur Jules Cotard', presented on 30 April 1894 to
   the Société Médico-Psychologique, Paris, p. 32.

10 J. Cotard, 'Du délire des négations I and II', *Archives de Neurologie*, IV
   (1882), pp. 152–70 and pp. 282–96.

11 Cotard, 'Délire des négations II', p. 289.

12 Cotard, 'Délire des négations I', pp. 153–4.

13 Cotard, 'Délire hypochondriaque', p. 172.

14 Ibid., p. 171.

15 G. E. Berrios and R. Luque, 'Cotard's Syndrome: Analysis of 100 Cases',
   Acta Psychiatrica Scandinavica, XCI/3 (1995), pp. 185–8.

16 S.N.C. Kudlur, S. George and M. Jaimon, 'An Overview of the
   Neurological Correlates of Cotard Syndrome', *European Journal of
   Psychiatry*, XXI/2 (2007), pp. 99–116.

17 B. G. Teixeira, A. F. Araújo and J. F. Perestrelo, 'Cotard's Syndrome: Two
   Cases of Self-Starvation', *Psilogos*, XIII/1 (2015), pp. 124–32.

18 J. Ramirez-Bermudez et al., 'Cotard Syndrome in Neurological and
   Psychiatric Patients', *Journal of Neuropsychiatry and Clinical Neurosciences*,
   XXII/4 (2010), pp. 409–16.

19 D. Cohen and A. Consoli, 'Production of Supernatural Beliefs during
   Cotard's Syndrome, a Rare Psychotic Depression', *Behavioral and Brain
   Sciences*, XXIX/5 (2006), pp. 468–70.

20 C. Tomasetti et al., 'The "Dead Man Walking" Disorder: An Update
   on Cotard's Syndrome', *International Review of Psychiatry*, XXXII/5–6
   (2020), pp. 500–509.

21 C. G. Huber and A. Agorastos, 'We Are All Zombies Anyway:
   Aggression in Cotard's Syndrome', *Journal of Neuropsychiatry and Clinical
   Neurosciences*, XXIV/3 (2012), E21.

22 Kudlur et al., 'Overview', p. 113.

23 A. Helldén et al., 'Death Delusion', *British Medical Journal*,
   CCCXXXV/7633 (2007), p. 1305.

24 A. Sahoo and K. A. Josephs, 'A Neuropsychiatric Analysis of the Cotard
   Delusion', *Journal of Neuropsychiatry and Clinical Neurosciences*, XXX/1
   (2018), pp. 58–65: p. 60.

25 H. Debruyne et al., 'Cotard's Syndrome', *Mind and Brain: The Journal of
   Psychiatry*, II/1 (2011), pp. 67–72.

26 D. Heller-Roazen, *The Inner Touch: Archaeology of a Sensation* (New York,
   2007).

27 J. Séglas, *Le Délire des negations: séméiologie et diagnostic* (Paris, 1897),
    p. 105.

28 J. Capgras and J. Reboul-Lachaux, 'L'illusion des "sosies" dans un délire
    systématisé chronique', *Bulletin de la Société de médicine mentale*, 11 (1923),
    pp. 6–16.

29 G. Blount, 'Letter to the Editor', *Nebraska Medical Journal*, LXXI/6
    (June 1986), p. 207.

30 J. L.C. Schroeder van der Kolk, *Handboek van de pathologie en therapie
    der krankzinnigheid* (Utrecht, 1863).

31 Ibid., p. 230.

32 S. Wright, A. W. Young and D. J. Hellawell, 'Sequential Cotard and
    Capgras Delusions', *British Journal of Clinical Psychology*, XXXII/3 (1993),
    pp. 345–9.

33 A. W. Young et al., 'Cotard Delusion after Brain Injury', *Psychological
    Medicine*, XXII/3 (1992), pp. 799–804.

34 H. D. Ellis and A. W. Young, 'Accounting for Delusional
    Misidentifications', *British Journal of Psychiatry*, CLVII/2 (1990),
    pp. 239–48.

35 A. W. Young, K. M. Leafhead and T. K. Szulecka, 'The Capgras and
    Cotard Delusions', *Psychopathology*, XXVII/3–5 (1994), pp. 226–31.

36 V. Charland-Verville et al., 'Brain Dead Yet Mind Alive: A Positron
    Emission Tomography Case Study of Brain Metabolism in Cotard's
    Syndrome', *Cortex*, XLIX/7 (2013), pp. 1997–9.

37 The article was published in June 2013. Helen Thomson then published
    the book *Unthinkable: An Extraordinary Journey through the World's
    Strangest Brains* (London, 2018).

38 Ibid., p. 200.

39 Ibid., p. 204.

40 Ibid.

41 J-P. Sartre, *The Wall and Other Stories*, trans. Andrew Brown (Richmond,
    2005).

42 S. Freud, 'Zeitgemäßes über Krieg und Tod', *Imago: Zeitschrift für
    Anwendung der Psychoanalyse auf die Geisteswissenschaften*, vol. IV
    (Vienna, 1915), pp. 1–21: p. 12, in *Reflections on War and Death*, trans.
    A. A. Brill and Alfred B. Kuttner (New York, 1918).

43 Ibid.

THREE The Murder of the Widow Van Sandbrink

1 I. Matthey, '"Genacht geliefden! Voor eeuwig genacht!" Lucas Kier van Ootmarsum en het einde der tijden' ['"Goodnight, dear ones! For ever, goodnight!" Lucas Kier van Ootmarsum and the End of Days'], *Flehite, Historisch Jaarboek voor Amersfoort en omstreken*, IX (2008), pp. 30–53. All the quotations (originally in Dutch) and much of the historical data in this chapter are taken from this work.

2 I. O'Donnell, R. Farmer and J. Catalan, 'Suicide Notes', *British Journal of Psychiatry*, CLXIII/1 (1993), pp. 45–8.

3 The letter is quoted in full (in Dutch) in Matthey's article, pp. 52–3.

4 Hendrik Henzepeter, *Gebeurtenissen der eerste en laatste tijden, welke in betrekking staan tot het begin en einde der Groote Wereld-Revolutie* (Amsterdam, 1832), p. 43.

5 S. van Ruller, *Genade voor recht. Gratieverlening aan ter dood veroordeelden in Nederland, 1806–1870* (Amsterdam, 1987).

6 G. E. Berrios and R. Luque, 'Cotard's Syndrome: Analysis of 100 Cases', *Acta Psychiatrica Scandinavica*, XCI/3 (1995), pp. 185–8.

7 I. Hacking, 'Kinds of People: Moving Targets', *Proceedings of the British Academy of Science*, CLI (2007), pp. 285–318.

8 H. Franke, *Twee eeuwen gevangen. Misdaad en straf in Nederland* (Utrecht, 1990), p. 89.

FOUR Phantoms and Illusions

1 Anonymous, 'The Case of George Dedlow', *Atlantic Monthly*, XVIII/105 (1866), pp. 1–11.

2 S. Weir Mitchell, 'The Medical Department in the Civil War', *Journal of the American Medical Association*, LXII/19 (1914), pp. 1445–50: p. 1448.

3 L. Figg and J. Farrell-Beck, 'Amputation in the Civil War: Physical and Social Dimensions', *Journal of the History of Medicine and Allied Sciences*, XLVIII/4 (1993) pp. 454–75: p. 454.

4 D. J. Canale, 'Civil War Medicine from the Perspective of S. Weir Mitchell's "The Case of George Dedlow"', *Journal of the History of the Neurosciences*, XI/1 (2002), pp. 11–18: p. 17.

5 J. Bourke, 'The Art of Medicine. Silas Weir Mitchell's The Case of George Dedlow', *The Lancet*, CCCLXXIII/9672 (2009), pp. 1332–3: p. 1332.

6 S. Finger and M. P. Hustwit, 'Five Early Accounts of Phantom Limb in Context: Paré, Descartes, Lemos, Bell, and Mitchell', *Neurosurgery*, LII/3 (2003), pp. 675–85.

7 H. Mayhew, *London Labour and the London Poor* (Ware, 2008), p. 154.

8 S. Weir Mitchell, 'Phantom Limbs', *Lippincott's Magazine*, VIII/48 (1871), pp. 563–9.

9 Ibid., pp. 565–6.

10 S. Weir Mitchell, *Injuries of Nerves and Their Consequences* (Philadelphia, PA, 1872), p. 348. Quoted from the 1965 reprint, New York.

11 Mitchell, 'Phantom Limbs', pp. 566–7.

12 Mitchell, *Injuries*, p. 350.

13 Guéniot, 'D'une hallucination du toucher (hétérotopie subjective des extrémités) particulière à certains amputés', *Journal de physiologie de l'homme et des animaux*, IV (1861), pp. 416–30.

14 Mitchell, *Injuries*, p. 349.

15 R. F. Reilly, 'Medical and Surgical Care during the American Civil War, 1861–1865', *Baylor University Medical Center Proceedings*, XXIX/2 (2016), pp. 138–42: p. 138.

16 W. James, 'The Consciousness of Lost Limbs', *Proceedings of the American Society for Psychical Research*, I (1887), pp. 249–58.

17 The 'man in his seventies' may have been the father of William (and Henry) James, Henry James senior. As a boy of thirteen, he had bravely tried to stamp out a fire in a barn and been so badly burned that his leg had to be amputated.

18 D. P. Kuffler, 'Origins of Phantom Limb Pain', *Molecular Neurobiology*, LV/1 (2018), pp. 60–69; A. Pirowska et al., 'Phantom Phenomena and Body Scheme after Limb Amputation: A Literature Review', *Neurologia i Neurochirurgia Polska*, XLVIII/1 (2014), pp. 52–9.

19 L. Hope-Stone et al., 'Phantom Eye Syndrome: Patient Experiences after Enucleation for Uveal Melanoma', *Ophthalmology*, CXXII/8 (2015), pp. 1585–90.

20 R. Cohn, 'Phantom Vision', *Archives of Neurology*, XXV/5 (1971), pp. 468–71.

21 G. Grouios, 'Phantom Smelling', *Perceptual and Motor Skills*, XCIV/3 (2002), pp. 841–50.

22 N. J. Wade and S. Finger, 'Phantom Penis: Historical Dimensions', *Journal of the History of the Neurosciences*, XIX/4 (2010), pp. 299–312: p. 304.

23 A. Marshal, *The Morbid Anatomy of the Brain in Mania and Hydrophobia* (London, 1815), p. 223.

24 A. P. Heusner, 'Phantom Genitalia', *Transactions of the American Neurological Association*, 78 (1950), pp. 128–31.

25 C. M. Fisher, 'Phantom Erection after Amputation of Penis: Case Description and Review of the Literature on Phantoms', *Canadian Journal of Neurological Science*, XXVI/1 (1999), pp. 53–6.

26 A. Crone-Münzebrock, 'Zur Kenntnis des Phantomerlebnisses nach Penisamputation', *Zeitschrift für Urologie*, XLIV (1951), pp. 819–22.

27 V. S. Ramachandran and P. D. McGeoch, 'Phantom Penises in Transsexuals: Evidence of an Innate Gender-Specific Body Image in the Brain', *Journal of Consciousness Studies*, XV/1 (2008), pp. 5–16.

28 P. U. Dijkstra, J. S. Rietman and J.H.B. Geertzen, 'Phantom Breast Sensations and Phantom Breast Pain: A 2-Year Prospective Study and a Methodological Analysis of Literature', *European Journal of Pain*, XI/1 (2007), pp. 99–108.

29 T. Mulder et al., 'Born to Adapt, But Not in Your Dreams', *Consciousness and Cognition*, XVII/4 (2008), pp. 1266–71.

30 C. S. Hurovitz et al., 'The Dreams of Blind Men and Women: A Replication and Extension of Previous Findings', *Dreaming*, IX (1999), pp. 183–93.

31 R. Bartholow, 'Experimental Investigations into the Functions of the Human Brain', *American Journal of Medical Science*, LXVII (1874), pp. 305–13.

32 Ibid., p. 309.

33 Ibid., p. 310.

34 W. Penfield and E. Boldrey, 'Somatic Motor and Sensory Representation in the Cerebral Cortex of Man as Studied by Electrical Stimulation', *Brain*, LX/4 (1937), pp. 389–443.

35 L. J. Harris and J. B. Almerigi, 'Probing the Human Brain with Stimulating Electrodes: The Story of Roberts Bartholow's (1874) Experiment on Mary Rafferty', *Brain and Cognition*, LXX/1 (2009), pp. 92–115: p. 108.

36 C. Holden, 'Monkey Euthanasia Stalled by Activists', *Science*, CCXLIV/4911 (1989), p. 1437.

37 T. P. Pons et al., 'Massive Cortical Reorganization after Sensory Deafferentation in Adult Macaques', *Science*, CCLII/5014 (1991), pp. 1857–60.

38 V. S. Ramachandran, M. Stewart and D. C. Rogers-Ramachandran, 'Perceptual Correlates of Massive Cortical Reorganization', *NeuroReport*, III/7 (1992), pp. 583–6.

39 V. S. Ramachandran and S. Blakeslee, *Phantoms in the Brain* (New York, 1998), pp. 35–6.

40 V. S. Ramachandran and D. C. Rogers-Ramachandran, 'Synaesthesia in Phantom Limbs Induced with Mirrors', *Proceedings of the Royal Society: Biological Sciences*, CCLXIII/1369 (1996), pp. 377–86.

41 Ibid., p. 381.

42 Ibid., p. 382.

43 Mitchell, 'Medical Department', p. 1448.

44 L. M. Hilti and P. Brugger, 'Incarnation and Animation: Physical versus Representational Deficits of Body Integrity', *Experimental Brain Research*, CCIV/3 (2010), pp. 315–26: p. 318.

45 M. Botvinick and J. Cohen, 'Rubber Hands "Feel" Touch that Eyes See', *Nature*, CCCXCI/756 (1998), p. 756.

46 K. Carrie Armel and V. S. Ramachandran, 'Projecting Sensations to External Objects: Evidence from Skin Conductance Response', *Proceedings of the Royal Society: Biological Sciences*, CCLXX/1523 (2003), pp. 1499–506.

47 O. Sacks, *A Leg to Stand On* (New York, 1984).

48 Ramachandran and Blakeslee, *Phantoms*, p. 58.

FIVE Whole at Last

1 R. C. Smith, 'Body Integrity Identity Disorder: The Surgeon's Perspective', in *Body Integrity Identity Disorder: Psychological, Neurobiological, Ethical and Legal Aspects*, ed. A. Stirn, A. Thiel and S. Oddo (Lengerich, 2009), pp. 41–8.

2 BBC Horizon, 'Complete Obsession', 17 February 2000. For the transcript, see www.bbc.co.uk.

3 C. Dyer, 'Surgeon Amputated Healthy Legs', *British Medical Journal*, CCCXX/7231 (5 February 2000), p. 332.

4 M. B. First, 'Desire for Amputation of a Limb: Paraphilia, Psychosis, or a New Type of Identity Disorder', *Psychological Medicine*, XXXV/6 (2005), pp. 919–28.

5 S. Noll and E. Kasten, 'Body Integrity Identity Disorder (BIID): How Satisfied Are Successful Wannabes', *Psychology and Behavioral Sciences*, III/6 (2014), pp. 222–32.

6 S. C. Schlozman, 'Upper-Extremity Self-Amputation and Replantation: 2 Case Reports and a Review of the Literature', *Journal of Clinical Psychiatry*, LIX/12 (1998), pp. 681–6.

7 Nelson, 'Living a Life with BIID', in *Body Integrity Identity Disorder*, ed. Stirn, Thiel and Oddo, pp. 82–7.

8 Ibid., p. 84.

9 J. Money, R. Jobaris and G. Furth, 'Apotemnophilia: Two Cases of Self-Demand Amputation as Paraphilia', *Journal of Sex Research*, XIII/2 (1977), pp. 115–25.

10 Ibid., p. 117.

11 R. von Krafft-Ebing, *Psychopathia Sexualis* (Stuttgart, 1886), quoted in *Psychopathia sexualis: A Medico-Forensic Study* (New York, 1939), pp. 236–7.

12 E. Kasten and F. Spithaler, 'Body Integrity Identity Disorder: Personality Profiles and Investigation of Motives', in *Body Integrity Identity Disorder*, ed. Stirn, Thiel and Oddo, p. 26.

13 O. Sacks, *A Leg to Stand On* (New York, 1984).

14 D. Vitacco, L. Hilti and P. Brugger, 'Negative Phantom Limbs? A Neurological Account of Body Integrity Identity Disorder', in *Body Integrity Identity Disorder*, ed. Stirn, Thiel and Oddo, pp. 201–10.

15 D. Brang, P. D. McGeoch and V. S. Ramachandran, 'Apotemnophilia: A Neurological Disorder', *NeuroReport*, XIX/13 (2008), pp. 1305–6.

16 G. Saetta et al., 'Neural Correlates of Body Integrity Dysphoria', *Current Biology*, XXX/11 (2020), pp. 2191–5.

17 M. Gheen, 'Clear Definitions and Scientific Understanding: Thoughts of an Academic Physician with BIID', in *Body Integrity Identity Disorder*, ed. Stirn, Thiel and Oddo, pp. 94–102.

18 B. Berger et al., 'Nonpsychotic, Nonparaphilic Self-Amputation and the Internet', *Comprehensive Psychiatry*, XLVI/5 (2005), pp. 380–83: p. 382.

19 I. Hacking, *Mad Travelers: Reflections on the Reality of Transient Mental Illnesses* (Charlottesville, VA, 1998).

20 C. Elliott, 'A New Way to Be Mad', *The Atlantic*, CCLXXXVI/6 (2000), pp. 72–84.

21 R. Dotinga, 'Out on a Limb', *Salon*, www.salon.com, 29 August 2000.

22 L. M. Hilti and P. Brugger, 'Incarnation and Animation: Physical versus Representational Deficits of Body Integrity', *Experimental Brain Research*, CCIV/3 (2010), pp. 315–26: p. 323.

23 J. Johnston and C. Elliott, 'Healthy Limb Amputation: Ethical and Legal Aspects', *Clinical Medicine*, II/5 (2002), pp. 431–5: p. 431.

24 P. Sue, *Anecdotes historiques, littéraires et critiques sur la médecine, la chirurgie, et la pharmacie* (Brussels, 1789).

25 Ibid., p. 223.

26 Hilti and Brugger, 'Incarnation', p. 321.

SIX Grief Hallucinations

1 W. F. Hermans, *Boze brieven van Bijkaart* (Amsterdam, 1977), p. 121.

2 M. Lundorff et al., 'Prevalence of Prolonged Grief Disorder in Adult Bereavement: A Systematic Review and Meta-Analysis', *Journal of Affective Disorders*, CCXII (2017), pp. 138–49.

3 W. Dewi Rees, 'The Hallucinations of Widowhood', *British Medical Journal*, IV/5778 (1971), pp. 37–41.

4 P. R. Olson et al., 'Hallucinations of Widowhood', *Journal of the American Geriatrics Society*, XXXIII/8 (1985), pp. 543–7.

5 K. S. Kamp et al., 'Bereavement Hallucinations after the Loss of a Spouse: Associations with Psychopathological Measures, Personality and Coping Style', *Death Studies*, XLIII/4 (2019), pp. 260–69.

6 C. M. Parkes, 'Psycho-Social Transitions: Comparison between Reactions to Loss of a Limb and Loss of a Spouse', *British Journal of Psychiatry*, CXXVII/3 (1975), pp. 204–10.

7 P. Auster, *Baumgartner* (London, 2023), p. 28.

8 Parkes, 'Transitions', p. 206.

9 Ibid., p. 207.

10 Ibid., p. 206.

11 Auster, *Baumgartner*, p. 30.

12 E. Earnest, *S. Weir Mitchell: Novelist and Physician* (Philadelphia, PA, 1950), quoted in A. D'Agostino, A. Castelnovo and S. Scarone, 'Non-Pathological Associations – Sleep and Dreams, Deprivation and Bereavement', in *The Neuroscience of Visual Hallucinations*, ed. D. Collerton, E. P. Mosimann and E. Perry (Hoboken, NJ, 2015), pp. 59–89: p. 75.

13 J. M. Schneck, 'Visual Hallucinations as Grief Reaction without the Charles Bonnet Syndrome', *New York State Journal of Medicine*, XC/4 (1990), pp. 216–17: p. 216.

14 J. Smith and E. V. Dunn, 'Ghosts: Their Appearance during Bereavement', *Canadian Family Physician*, XXIII (1977), pp. 121–2.

15 C. Baethge, 'Grief Hallucinations: True or Pseudo? Serious or Not? An Inquiry into Psychopathological and Clinical Features of a Common Phenomenon', *Psychopathology*, XXXV/5 (2002), pp. 296–302.

16 J. Hayes and I. Leudar, 'Experiences of Continued Presence: On the Practical Consequences of "Hallucinations" in Bereavement', *Psychology and Psychotherapy: Theory, Research and Practice*, LXXXIX/2 (2016), pp. 194–210.

17   Ibid., p. 202.

18   Ibid.

19   C. Valentine, 'Identity and Post-Mortem Relationships in the Narratives of British and Japanese Mourners', *Sociological Review*, LXI/2 (2013), pp. 383–401: p. 388.

20   Ibid.

21   Ibid.

22   Ibid.

23   C. Bonnet, *Essai analytique sur les facultés de l'âme* (Copenhagen, 1760).

24   D. Draaisma, *Disturbances of the Mind*, trans. B. Fasting (Cambridge, 2009).

25   G. de Morsier, 'Les Automatismes visuels (hallucinations visuelles rétro-chiasmatiques)', *Schweizerische Medizinische Wochenschrift*, 29 (1936), pp. 700–703.

26   R. J. Teunisse, *Concealed Perceptions: An Explorative Study of the Charles Bonnet Syndrome* (Nijmegen, 1998).

27   Draaisma, *Disturbances of the Mind*, p. 29.

28   E. Canetti, *Das Augenspiel* (Frankfurt, 1988). English translation of excerpts by Jane Hedley-Prôle.

29   Ibid.

ACKNOWLEDGEMENTS

In researching the chapter on the three Christs of Ypsilanti, I profited greatly from a discussion with Ruth Benschop; it is her I have to thank for the final sentence.

I am indebted to the historian Ignaz Matthey for much of the material underpinning the chapter on the murder of the widow Van Sandbrink of Amersfoort. He not only wrote a beautifully documented study of the murder but alerted me to an early case of Cotard's, noted in 1688 by Jan Six and described by Geert Mak in his book about the Six family. Matthey was generous in sharing his material. Sadly, he passed away in June 2022.

Comments by Paul Andreoli and Liesbeth Baas on the chapter on grief hallucinations were greatly helpful.

All chapters, from the earliest drafts to the final versions, benefited from the sharp eyes and comments of Anne Boomsma. And Patrick Everard did what he has always done: helped me raise my manuscript to the level of a book he deemed worth publishing.

I thank the staff of Reaktion Books for their meticulous fact-checking and research on illustrations. My editor Amy Salter offered tactful guidance in crafting a final manuscript that is now without a doubt a better version of what had landed on her desk initially.

I am particularly grateful to Jane Hedley-Prôle, who did so much more than translating my book. Along the way, she did a careful check of the evidence, which resulted in several amendments. With a keen eye for coherence and style, she came up with many suggestions for improvements, which I thankfully accepted. Working with Jane was a delight – pure and simple.

PHOTO ACKNOWLEDGEMENTS

The author and publishers wish to express their thanks to the sources listed below for illustrative material and/or permission to reproduce it. Some locations of works are also given below, in the interest of brevity:

Ann Arbor District Library, MI: pp. 24, 25; from R. Bartholow, 'Medical Electricity and Medical Electrical Apparatus. The Electrical Room of the Good Samaritan Hospital', *The Clinic*, II/8 (24 February 1872), photo courtesy the Henry R. Winkler Center for the History of the Health Professions, Donald C. Harrison Health Sciences Library, University of Cincinnati, OH: p. 130; Bibliothèque Charcot, Sorbonne Université, Paris: p. 68; Leiden University Libraries: p. 94; Library of Congress, Prints and Photographs Division, Washington, DC: pp. 110, 120; Musée Carnavalet, Histoire de Paris: pp. 13, 14; National Library of Medicine, Bethesda, MD: p. 114; National Museum of American History, Smithsonian Institution, Warshaw Collection of Business Americana, Archives Center, Washington, DC: p. 121; from Wilder Penfield and Theodore Rasmussen, *The Cerebral Cortex of Man: A Clinical Study of Localization of Function* (New York, 1950): p. 133; from V. S. Ramachandran and William Hirstein, 'The Perception of Phantom Limbs', *Brain*, CXXI/9 (September 1998), p. 1621: p. 141; Washington State University Libraries, Manuscripts, Archives, and Special Collections (MASC), Pullman: p. 28.